T[illegible]ket

R. G. HIBBERD
B.Sc., M.I.E.E., Sen.M.I.E.E.E.

Newnes Technical Books

Newnes Technical Books
is an imprint of the Butterworth Group
which has principal offices in
London, Sydney, Toronto, Wellington, Durban and Boston

First published 1965 by George Newnes Ltd
Reprinted 1968, 1971 (by Newnes–Butterworth), 1973, 1976
(by Newnes Technical Books), 1981

ISBN 0 408 00254 9

Typeset by Butterworths Litho Preparation Department
Printed in England by Butler & Tanner Ltd., Frome & London

PREFACE

When the publishers approached me to write this book, they stated: "The books in this series are intended mainly for students and technicians, the treatment should be kept free of mathematics and emphasis should be placed on practical considerations."

Throughout the preparation of the manuscript, I have borne this in mind, and have generally attempted to explain topics in words and description. This, at times, has necessitated some simplification and approximation, but I hope that as a result of reading the book students and technicians will be encouraged to study further some of the more advanced books on the theory and specialised applications of transistors.

I gratefully acknowledge help from colleagues in Texas Instruments Limited (England) and Texas Instruments Incorporated (U.S.A.) especially with regard to many of the practical circuits.

R.G.H.

CONTENTS

1
HISTORICAL INTRODUCTION

To trace the trend of events leading up to the invention of the transistor it is convenient to recall that Ohm's Law, relating to linear electrical conduction in solids, was first stated in the year 1827. It was not until 1870 that electrical non-linearity, or rectifying action, was observed in metallic sulphides such as galena and pyrites. Early in the twentieth century the advent of radio telegraphy led to further exploration of the sulphides as radio detectors, and during the early days of radio broadcasting, in the 1920–25 period, there was widespread use of the crystal detector in crystal receivers. These detectors in general used a small piece of galena with a metal "catswhisker" contact.

Silicon crystal mixer diode

The thermionic valve had been originated back in 1904, and by 1930 was being widely used in radio receivers. Developments during the next decade were rapid, and electronic equipment using thermionic valves was coming into common use by the start of World War II. Progress continued at even higher tempo during the war.

The development of high definition radar using signal frequencies of 3,000 and later 10,000 Mc/s brought the need for a detector with better performance than was possible with thermionic diodes, and attention was again focused on the crystal detector in view of its low capacitance and low forward resistance. Further work resulted in the silicon crystal diode, consisting of a small piece of silicon with a pointed tungsten wire pressing on to its surface. This diode was used mainly as a "mixer" diode for superheterodyne reception with a local oscillator, but there were also several uses as a straight detector of signal frequency particularly for monitoring and measurement.

Germanium point contact diode

Towards the end of the war, during experiments to obtain an improved mixer diode, it was discovered that relatively high reverse voltage characteristics could be obtained by using germanium. By now it was becoming apparent that the success of future work on these devices would depend on a fuller understanding of the semiconductor materials, particularly germanium and silicon, and a better knowledge of how to control their characteristics. While more fundamental work of this nature was being done, the germanium point contact diode was engineered into a reasonably satisfactory device. It had good general purpose characteristics for low and medium frequency applications—a forward resistance of a few hundred ohms, a reverse resistance about ten thousand ohms with allowable forward current of 20 milliamperes and reverse voltage up to 100 volts. It soon began to receive wide acceptance in electronic circuits as a detector and a limiter in place of the thermionic diode previously used for these purposes. But at this time the operation of the diode was not well understood, and its design and manufacture was largely empirical.

Point contact transistor

Meanwhile, work to establish sound theoretical understanding of rectification was proceeding. During the course of such work at the Bell Telephone Laboratories in 1948 scientists were investigating the surface potentials around a point contact by using a second contact to probe the field. When the spacing between the two points was very small, one or two thousandths of an inch, they observed that a change of current in the main contact influenced the current flowing in the other. This observation led to the point contact transistor, which consists of a piece of flat germanium with two pointed metal wires pressing on to it about one-thousandth of an inch apart. Each contact alone has a rectifying characteristic. The first, called the emitter, is biased in the forward direction, and so has a low resistance, while the second, called the collector, is biased in the reverse direction giving a high resistance (forward and reverse biasing are explained in the next chapter). A change of the forward current through the low resistance emitter produces a similar change of current in the high resistance collector and so power amplification is obtained. The concept of "transferring" a current from a low resistance to a high resistance circuit is one version of the origin of the name transistor—*TRANS*fer-res*ISTOR*.

This was the first time that amplification had been observed in a solid state device and the implications were immediately apparent. The new device was much smaller than a thermionic valve and needed no heater supply. It offered reduced size, weight and power consumption of electronic equipment. Moreover, at this time, the development of the thermionic valve had reached a state of saturation, and this gave added interest to the new alternative.

Junction transistor

It was soon realised that the point contact transistor suffered from several limitations. It was difficult to manufacture specimens with the same characteristics, the cost was high, it lacked ruggedness and had a high noise factor. Nevertheless, it had already served its purpose by demonstrating the feasibility of solid state amplification and had encouraged intensive programmes of research, the first outcome of which was the junction transistor, also invented at the Bell Telephone Laboratories. Instead of using two metal wire contacts at the surface of the semiconductor, two rectifying junctions are formed inside the solid, the distance between the junctions again being small—less than one-thousandth of an inch. During the period 1950 to the present time, the development of the junction transistor has passed through a number of phases as new technologies have been developed, each phase resulting in a new and better type of transistor—grown junction, alloy, surface barrier, alloy diffused, diffused mesa, epitaxial diffused mesa, planar, epitaxial planar, diffused field effect (metal-oxide silicon transistor) and so on.

It is this family of transistors that forms the subject of this book.

2
BASIC PRINCIPLES

To understand the operation of the transistor—how it is that currents can flow in a piece of solid crystal in such a way as to result in amplification—it is necessary to start with the atomic structure of the crystal. In a much simplified picture, the atom consists of a central nucleus containing positively charged protons, around which rotate a number of negatively charged electrons. The negative charge of the electron is exactly equal and opposite to the positive charge of the proton. In a neutral atom, the number of electrons orbiting the nucleus is equal to the number of positively charged protons in the nucleus.

Different elements have different numbers of electrons, and these electrons do not all rotate at the same distance from the nucleus, but in several separate rings. For instance, germanium has a total of 32 electrons located in 4 rings, and silicon has 14 electrons in 3 rings. The electrons located in the rings close to the nucleus are tightly bound, but those in the outer ring are comparatively loosely bound and can, under certain circumstances, be dislodged from the atom. The properties of an element are largely determined by the conditions existing in its outer ring of electrons. This outer ring is called the *valence ring* and the electrons in it the *valence electrons*. These are normally the only electrons entering into chemical reactions and so it is convenient to imagine a simplified picture of the atom as a central nucleus with the valence electrons in one ring around it.

Conductors and insulators

Materials having the valence electrons very loosely bound to the atom are good conductors of electricity. The obvious example is copper, whose valence electrons are so loosely bound that the thermal energy present at room temperature is sufficient to cause approximately one electron to become detached from each atom. In an isolated bar of copper, these free electons will move about in random fashion. If a voltage is applied across the ends of the bar, the electrons will be attracted towards the positive end and a current will flow. The electrons will flow out of the bar to the battery at the positive end, and an equal number will flow from the battery into the negative end of the bar, so that the number of electrons in the bar is always the same. An important point to remember is that when an electron becomes detached from an atom the atom is left with a net positive charge, and although there are many free negative electrons in the bar it does not become electrically charged as there are an equal number of positively charged atoms.

Thus we see that in metals, which are good conductors of electricity, there normally exists a number of electrons which are free to move when an electrical potential is applied.

Insulators are materials in which the valence electrons are tightly bound to the nucleus, and need a large amount of energy to release them. There will normally be only a very small number of free electrons in the material, and so practically no current will flow when a voltage is applied.

Semiconductors

Between the two major categories of conductors and insulators is the class of materials called semiconductors. As the name implies, a semiconductor is a material with resistivity roughly midway between conductors and insulators. However, a semiconductor is not just a poor conductor; it has two other important properties. First, its resistance normally decreases with increase of temperature, as opposed to conductors, such as metals, the resistance of which increases slightly with temperature. Secondly, flow of current in a semiconductor can be by two mechanisms, by a flow of negative electrons as in the case of conductors, or by a process known as a flow of positive holes in the opposite direction. This concept of hole flow is most important in the understanding of semiconductors and will be discussed in some detail later in this chapter.

Crystal structure

When the atoms of an element are close together, as in the case of a solid, they tend to align themselves in a uniform pattern and form what is called a crystalline structure. The exact form of the crystal depends on the number of valence

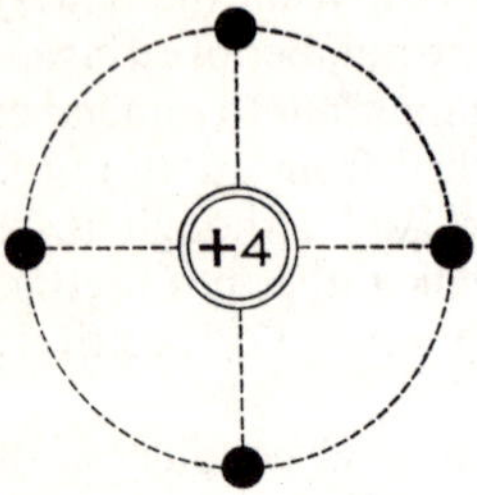

Fig, 2.1.—*Representation of a germanium or silicon atom.*

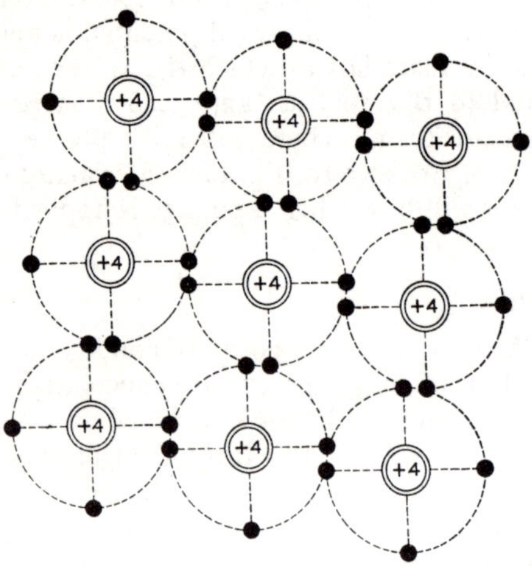

Fig. 2.2.—*Representation of a germanium or silicon crystal.*

electrons and how they can form bonds between neighbouring atoms. Germanium and silicon atoms have four valence electrons in their outer rings, and they both form the same basic structure in which every atom is equidistant from four other atoms, and each valence electron of one atom forms a pair, called a covalent bond, with one of another atom. The structure is, of course, three dimensional, but it is convenient to draw it in two dimensions as shown in Fig. 2.1, which is a representation of a germanium or silicon atom, and Fig. 2.2 which illustrates the structure of a perfect crystal. The large circles in the centre represent the nucleus, and the figure in it the number of positively charged protons in the nucleus. The small black circles each represent one electron.

Electrons and holes

In the perfect crystal, as in Fig. 2.2, all the covalent bonds are complete; no electrons are dislodged from their atoms. In this state, even germanium and silicon would be insulators. However, consider what will happen as the temperature of such a crystal is increased from absolute zero. As the temperature is increased, the energy of each atom is increased, and a valence electron may acquire sufficient energy to break away from its parent atom. Such electrons are now free, and if a voltage is applied they can move to form a current flow as in the case of conductors.

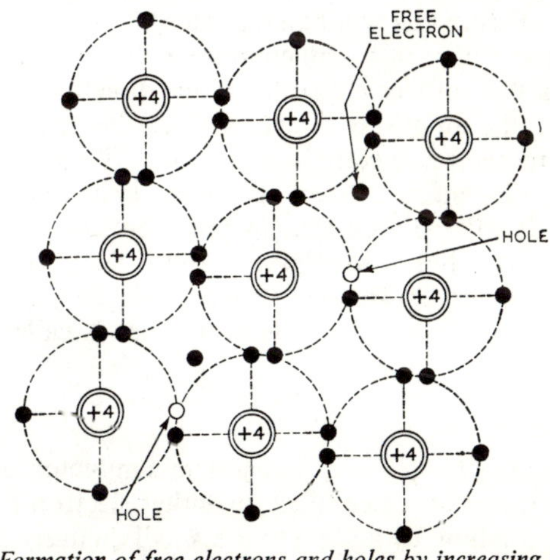

Fig. 2.3.—*Formation of free electrons and holes by increasing temperature.*

In breaking away from the atom, the electron leaves behind an empty space, called a *hole*, as shown in Fig. 2.3. The presence of holes can lead to conduction in a manner different from that of free electrons. A hole which has been created in an atom by an electron breaking free will effectively have a net positive charge. A negative electron which has just broken free from a neighbouring atom can jump into the hole and neutralise the charge. However, in breaking away from the neighbouring atom the electron has left a new positively charged hole behind it. Thus we can say that the hole has moved from the first atom to the second. Then, an electron from a third atom may fill the second hole, and the process repeats from atom to atom. If a voltage is applied, the electrons will move towards the positive end in a succession of jumps, and holes will appear to move in the opposite direction towards the negative end. Although in fact only electrons are moving in the crystal, it is convenient to consider the hole as a positively charged entity, capable of moving in a similar manner but in the opposite direction. Since the hole moves in a series of short movements, it will move slower than a free electron, which can move steadily along between the atoms. This fact can be important in the design of transistors as will be seen later.

Thus current flow in a semiconductor normally consists of two parts, a flow of free electrons in one direction, and a flow of holes in the opposite direction. As the temperature is increased, more free electrons and holes will be formed, so that there are more current carriers, and the resistance decreases. It will be seen that in pure semiconductors there must be one hole for every free electron, and so the total number of free electrons must be equal to the total number of holes. Such pure material is called *intrinsic* material.

Impurity conduction

It is possible to arrange that a semiconductor has a predominant single type of conduction, either electron conduction or hole conduction, by adding to it a small quantity of another element. If an element having five valence electrons is added, an excess of free electrons is created and electron conduction is obtained. An impurity having only three valence electrons will result in holes in the structure and will give hole conduction. These two conditions will be discussed in turn.

n-Type semiconductor

Consider what happens when an impurity having five electrons in the valence ring is added to otherwise pure germanium. The situation is illustrated in Fig. 2.4. The atoms

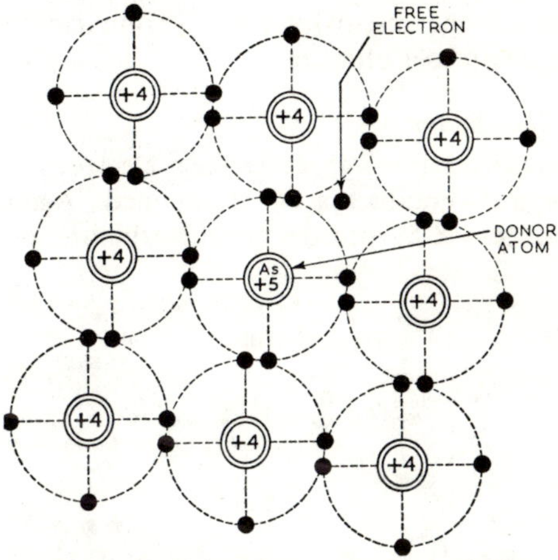

Fig. 2.4.—*Formation of free electrons by donor impurity atoms.*

of the impurity element will fit into the crystal structure as best they can. Four of the valence electrons will form covalent bonds with the valence electrons of neighbouring germanium atoms. The fifth valence electron of the impurity atom will tend to orbit around the atom, but it will be so loosely bound that at room temperature it is effectively free and available as a current carrier. Since the impurity atom now has only four valence electrons bonded to it, it is positively charged and is called a *positive ion.* The crystal as a whole remains neutral since for every positively charged impurity atom there is a free negative electron, but it is important to remember that while the electrons are free to move, the positively charged impurity atoms are fixed in the crystal lattice.

Typical impurities with a valency of five are arsenic, antimony and phosphorus. Since they donate free electrons to the semiconductor, they are called *donors*, and since the current carriers are negative electrons, the material is called *n-type* material (N for negative carriers). The addition of very small amounts of the impurity greatly changes the resistance of the semiconductor. A typical amount of impurity added to germanium for transistors is one part in a hundred million, and this will reduce the resistivity from 70 ohm cm for intrinsic germanium down to 1 ohm cm.

p-Type semiconductor

If an impurity having three valence electrons is added to germanium, a different situation is obtained. Again the atom of the impurity will fit into the crystal structure, but this time

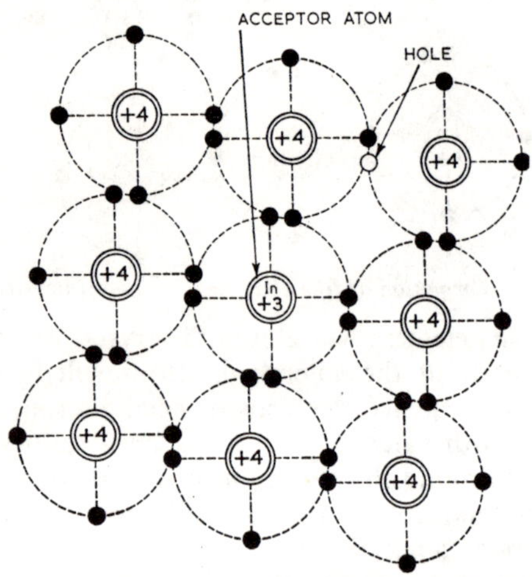

Fig. 2.5.—*Formation of holes by acceptor impurity.*

only three covalent bonds can be formed. In order to complete the fourth bond, the impurity atom will tend to attract an electron from a neighbouring germanium atom, giving rise to a hole in that atom. This situation is illustrated in Fig. 2.5.

Each atom of the impurity element will result in one hole in the germanium structure, and hole conduction will result, as described earlier. The stationary impurity atom, now having an extra electron has a net negative charge, and is a *negative ion*. Typical impurity elements with three valence electrons are indium, boron and aluminium. Since they accept an electron from the semiconductor, they are called *acceptors*, and since current flow is now by means of positively charged holes, the semiconductor material is called *p-type* (P for positive carrier).

Generation and recombination

As discussed earlier, if the temperature of a semiconductor is increased from absolute zero, electrons obtain more energy and some break away from their atoms, creating an equal number of free electrons and holes. These electrons and holes are said to be thermally generated. If the semiconductor is held at a given temperature—say at room temperature, holes and electrons will be continuously generated. It might seem that their concentrations would steadily increase. This however is not so. As a free electron moves around the crystal, it may encounter a hole. If it does, it will move into the hole and form a valence bond. Then the hole no longer exists, the electron is no longer free and both current carriers have disappeared. This process is called recombination. With the semiconductor at a constant temperature, the rate of recombination must be equal to the rate of generation since the material is in equilibrium.

With impurity conduction there is normally an excess of one type of carrier. This is called the *majority carrier*, and will be the electron in n-type material and the hole in p-type material.

Minority carrier recombination

If a number of holes are suddenly injected into an n-type semiconductor, as opposed to being continuously generated in it, they will gradually disappear as they recombine with free electrons, until they have all ceased to exist. Such holes present in n-type material are called *minority carriers*. The time for which they continue to exist before eventually recombining is called the *minority carrier lifetime*, and is an important property of the semiconductor material.

In a similar way, electrons injected into p-type semiconductor material are also minority carriers and will also disappear by recombination as they move into holes.

The operation of most transistors depends upon the movement of minority carriers in semiconductor material of opposite type. More will be said about this later.

The pn junction

A transition from p-type to n-type material in a continuous crystal of semiconductor is called a pn junction. Most semiconductor devices contain one or more pn junctions, and so it is very important to have a good understanding of its properties.

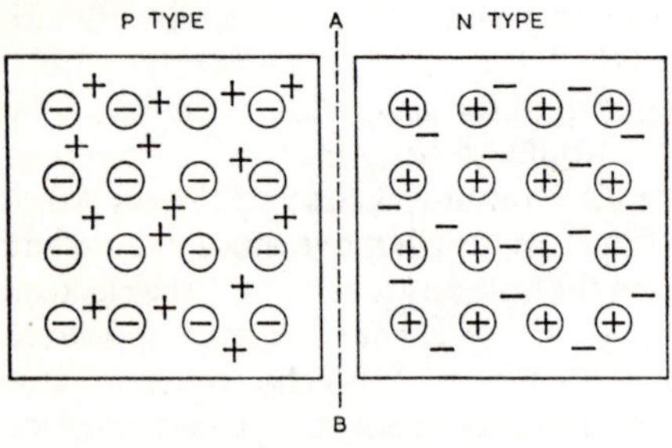

Fig. 2.6.—*Charges in separate p-type and n-type materials.*

Alone, a piece of n-type or p-type semiconductor is purely resistive, that is, reversing a battery connected across it will reverse the direction of current flow but will not affect the amplitude of the current. In contrast, a piece of semiconductor with a pn junction in it has rectifying properties—when a voltage is applied in one direction current flows, but if the voltage is reversed no current flows.

Consider two pieces of semiconductor material, one p-type with a certain density of holes, and the other n-type with a similar density of free electrons. In Fig. 2.6 to the right of the line AB is an n-type region, with fixed positively charged donor atoms and mobile negative electrons. It is important to remember that the positive donor atoms are fixed in the crystal structure, but the negative electrons are free to move under an applied field. Similarly to the left of the line AB is a p-type region with fixed negatively charged acceptor atoms and mobile positively charged holes. In the diagram only the donor and acceptor atoms and the electrons and holes are shown. The semiconductor atoms should be imagined as an orderly and continuous background over the whole material.

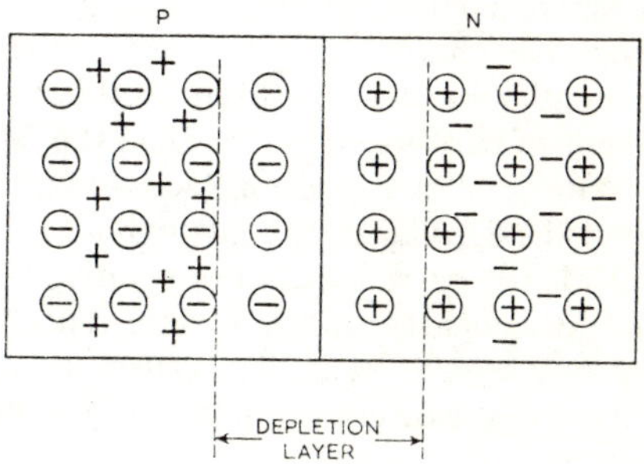

Fig. 2.7.—*Charges in an unbiased pn junction.*

In the p-type material, the acceptor atoms will normally be distributed regularly through the semiconductor and the holes, although free to move, will tend to be fairly evenly distributed also, as illustrated in Fig. 2.6. Similarly, the donor atoms and free electrons will be regularly distributed in the n-type material. Now, consider what happens if the two pieces suddenly become joined together with a continuous crystal structure as shown in Fig. 2.7. In the n-type region, there is a high concentration of electrons, while in the p-type region there is a low concentration of electrons. Consequently, electrons near the junction on the n side will tend to diffuse across the junction into the p side. As they move into the p-type material, they will recombine with holes which are present in high concentration, and they will move only a short distance into the p region before disappearing by recombination.

Similarly, the same situation occurs with the holes in the p side near the junction, which tend to diffuse across the junction in the opposite direction into the n side and recombine with electrons as they move in.

Now when the electrons diffuse over the junction into the p-type region, they leave behind the fixed positively charged donor atoms. Thus a net positive charge is built up on the n side of the junction. Similarly a net negative charge is established in the p side due to the remaining negatively charged acceptor atoms. Once these net charges have been established by the initial diffusion they discourage further diffusion, the positive charge on the n side now repelling holes and the negative charge on the p side repelling electrons. Thus a potential barrier has been established against further movement of carriers. This is called the *junction barrier.*

Outside this barrier on each side of the junction, the material is still neutral, i.e. the positive and negative charges balance. Inside the barrier there is a positive charge on the n side and a negative charge on the p side. This region is called the *space charge region.* It is also called the *depletion layer*, since the density of carriers in the region has been depleted.

pn Junction with forward voltage

Forward voltage is applied to a pn junction as shown in Fig. 2.8 (a) by connecting the positive terminal of a battery to the p side and the negative terminal to the n side. As the voltage is increased from zero, the free electrons in the n side and the holes in the p side are repelled towards the junction, the first effect being to neutralise some of the charged donor and acceptor atoms, and so reduce the potential barrier. So far no current flows. Further increase of voltage will eventually reduce the barrier to zero, and then the electrons and holes will move across the junction, the electrons eventually recombining with holes in the p region, and the holes recombining with electrons in the n region. Thus a current now flows across the junction, and this current increases rapidly with further increase of voltage. The nature of the current changes across the crystal. In the p region away from the junction, the current consists of a flow of holes, and in the n region away from the junction it consists of a flow of electrons. In the vicinity of the junction the current is a combination of hole and electron flow, the proportions varying with distance as shown in Fig. 2.8 (b).

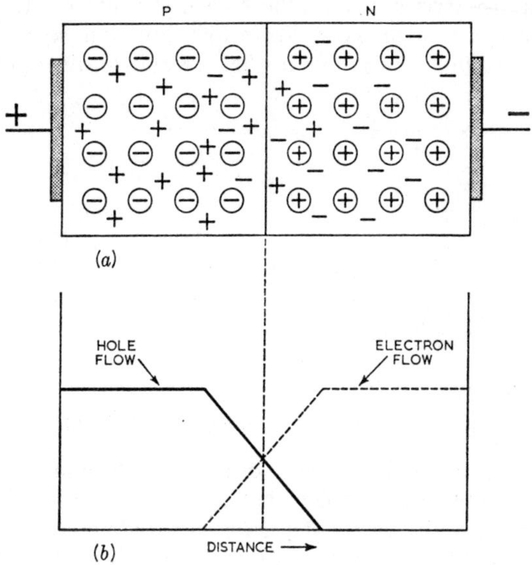

Fig. 2.8.—*(a) Charges in a forward biased pn junction. (b) Hole and electron flow in a forward biased junction.*

As the flow of holes in the p side approaches the junction it will start to encounter free electrons that have crossed over the junction from the n side. Recombination will start to occur and will build up as the holes proceed across the junction, until eventually, a short distance into the n-type region all the holes will have disappeared. Similarly the electron flow in the n region will decrease as it approaches the junction and fall to zero a short distance into the p region.

The important point to realise is that for a short distance on either side of the junction part of the current is carried by minority carriers which have crossed the junction, or as we say have been *injected* into a region of opposite conductivity type.

pn Junction with reverse voltage

If the voltage across the pn junction is reversed, the holes in the p-type region are attracted away from the junction towards the negative terminal, and the free electrons in the n-type region are attracted away from the junction towards the positive terminal. As shown in Fig. 2.9, this results in more of

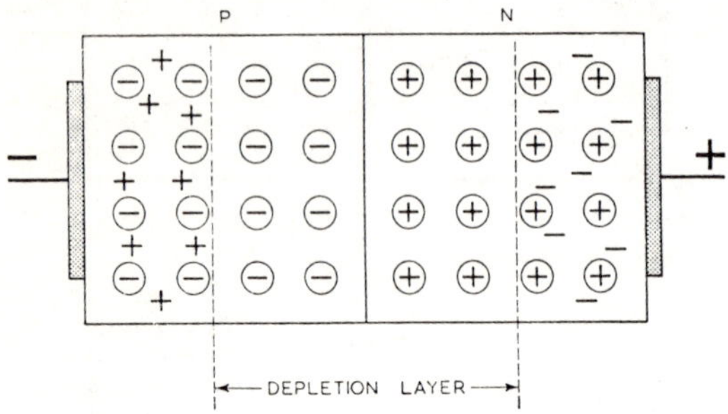

Fig. 2.9.—*Charges in a reverse biased pn junction.*

the fixed positively charged donor atoms and negatively charged acceptor atoms near the junction being left unneutralised and the width of the depletion layer is increased. In this condition no majority current can flow, but a very small current does flow by minority carriers. A certain number of holes and electrons will be thermally generated, and holes on the n side and electrons on the p side will be attracted across the junction to form a small current. The magnitude of this current will depend only on the rate of generation of the holes and electrons and this in turn depends only on the temperature. All of the carriers are attracted across the junction as they are generated and so increasing the voltage cannot increase the current. Thus this small reverse current remains substantially constant as the voltage is increased, and is called the *reverse saturation current*. It is important to remember that this saturation current increases with temperature, and we shall see later that if the temperature is increased enough the current will eventually become so large that it will present a limit to the operation of the pn junction.

Electrical characteristics of a pn junction

With the previous sections in mind, it is now possible to indicate the general form of the electrical characteristics of a pn junction. This is illustrated in Fig. 2.10. As the forward

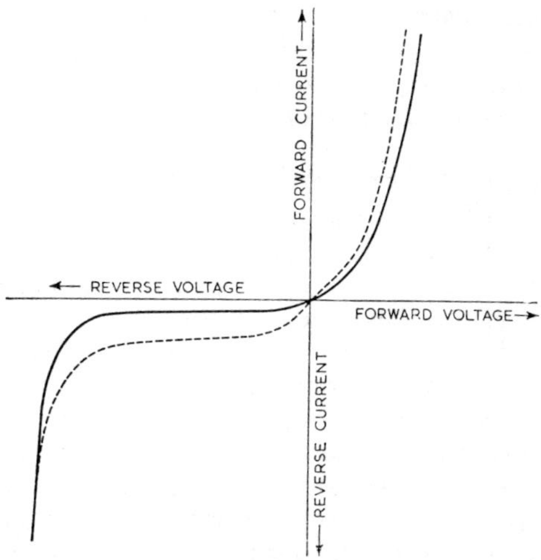

Fig. 2.10.—*Electrical characteristics of a pn junction.*

voltage is increased from zero, the forward current increases slowly at first until the junction barrier is overcome and then increases rapidly. The resistance of the forward biased junction is low. In the reverse direction, the current remains low and constant as the voltage is increased—the reverse resistance is very high. Eventually the voltage reaches a value called the *reverse breakdown voltage*. As the voltage is increased, the minority carriers constituting the reverse saturation current will move faster and eventually may be moving so fast that on colliding with more atoms they break covalent bonds and produce more carriers. These new carriers then accelerate and break the bonds of other atoms and so on. Thus a build-up of carriers is produced and what is known as *avalanche breakdown* occurs. The value of voltage at which this happens depends mainly on the initial concentration of carriers present, that is on the resistivity of the material.

Effect of temperature

It was mentioned above that the reverse saturation current increases with temperature. This is illustrated by the dotted curve in Fig. 2.10. The forward current also increases slightly with temperature since the total concentration of carriers is increased by thermal agitation.

Capacitance of a pn junction with reverse voltage

Another interesting characteristic of a reverse biased pn junction is that it resembles a capacitor. In the space charge region there are no mobile carriers and so the material is in effect an insulator. On either side are the n-type and p-type materials, which are conductors. Thus the conditions are similar to those in a conventional capacitor, which has two conducting electrodes separated by an insulator. The capacitance of a capacitor depends on the distance between the electrodes—as the distance is increased, the capacitance decreases and vice-versa. With the reverse biased junction, the distance between the conducting regions is effectively the width of the space charge layer, which, as indicated earlier, increases as the reverse voltage is increased.

Thus a reverse biased pn junction has the properties of a capacitor, the capacitance of which varies inversely with the reverse voltage.

Transistor action

Consider the hole flow in a pn junction with forward voltage biasing. It was shown earlier (Fig. 2.8) that holes are injected across the junction from the p region into the n region, and their density steadily decreases to zero as they flow on and recombine with free electrons. Now suppose that a second pn junction is arranged on the n side as shown in Fig. 2.11, at a point so close to the first junction that the hole density has only decreased slightly by recombination. If this second junction is biased in the reverse direction, it will normally have a high resistance and will oppose current flow. However, the situation it sees is abnormal in that the n-type region now contains a high concentration of positive holes which have come from the first p region. As these holes come under the influence of

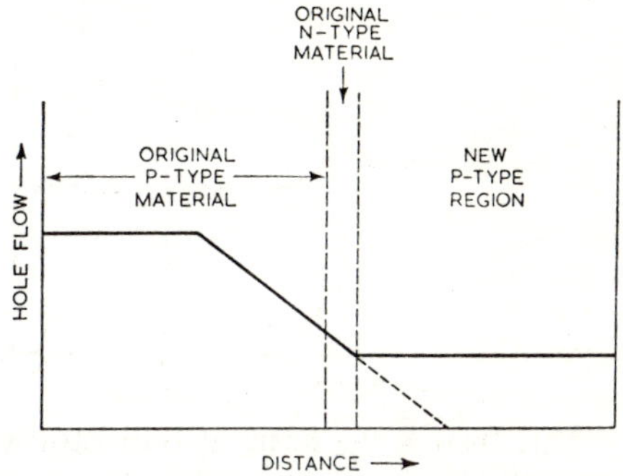

Fig. 2.11.—*Hole flow in a pnp structure (equal donor and acceptor concentrations).*

the space charge region of the second junction they will be attracted across the junction into the second p region. Once in the second p region, these holes will continue to flow as a normal majority current in p-type material as indicated by the horizontal line in Fig. 2.11.

Up to now, we have been considering the situation that exists when the acceptor concentration in the first p region and the donor concentration in the n region are approximately equal, and the current crossing the forward biased junction consists of one-half of a flow of holes in one direction and one-half of electrons in the other direction. For this case, the current in the second p region will be slightly less than half of the input current to the first p region. If, however, the concentration of holes in the first p region is made very large compared with the concentration of electrons in the n region, practically all of the current crossing the first junction will consist of hole flow, and then the current flowing in the second p region will be almost equal to the current flowing into the first p region as indicated in Fig. 2.12. The ratio of the hole current to the total current crossing the first junction is called the *injection efficiency* and can readily be made of the order of

99 per cent. The ratio of the hole current arriving at the second junction to the hole current injected into the n region is called the *transport efficiency* and can also be made to be 99 per cent.

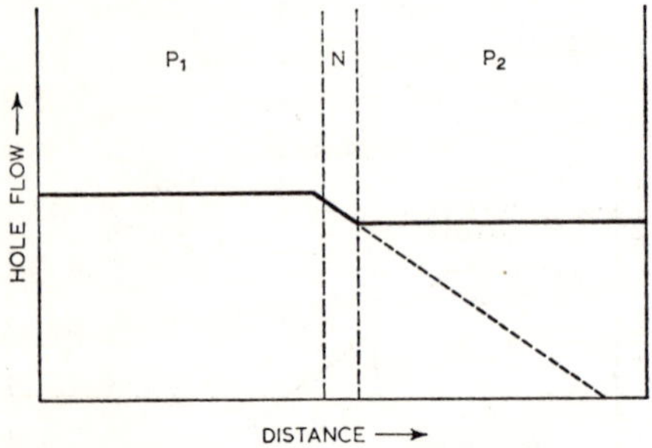

Fig. 2.12.—*Hole flow in a pnp structure with heavy acceptor concentration in p_1 region and light donor concentration in n region.*

This then is the basis of transistor action: it is possible to generate a current through a low resistance forward biased junction by applying a small voltage, and then to arrange that this current flows through an adjacent high resistance reverse biased junction, resulting in power amplification.

pnp Junction transistor

The concepts described above are embodied in the junction transistor which consists of two pn junctions formed back-to-back in a single piece of semiconductor and separated by a distance less than one-thousandth of an inch. It can be arranged either in pnp or npn form. Considering the pnp arrangement first, Fig. 2.13 shows a diagrammatic representation with associated nomenclature. The first p region, positively biased, is called the *emitter*, as it emits holes into the centre n region, which is called the *base*. The second p region is negatively biased, and is called the *collector* as it collects holes from the base region. The junction between emitter and base is called the *emitter junction*, and that between the collector and base the *collector junction*.

It has been shown that in passing through the base region the emitter current decreases by a small amount due to some holes recombining with electrons. To take the place of the electrons in the base lost in this way more must flow in from the base contact, such that the current so produced is equal to the difference between the emitter current and the collector current.

$$I_B = I_E - I_C$$

If a small change is made in the emitter current, the ratio of the change of collector current to the change of emitter current is called the *current transfer ratio* and is designated by the symbol α

$$\alpha = \frac{\Delta I_C}{\Delta I_E}$$

(the symbol Δ or δ indicating increase, i.e. ΔI_C increase in collector current).

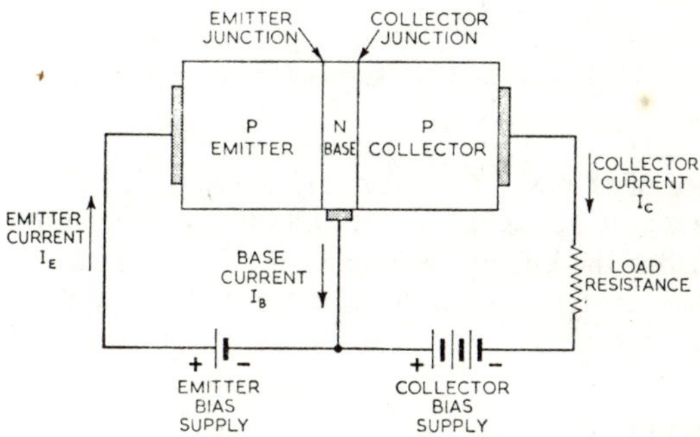

Fig. 2.13.—*pnp junction transistor.*

By suitable control in manufacture α can be made in the range 0·95 to 0·995.

Operation of the transistor as an amplifier can best be illustrated by considering a practical example. Consider the circuit in Fig. 2.14. Suitable d.c. conditions are first established by setting V_E such that a convenient emitter current flows—say 0·5 mA. This will result in a slightly lower collector current which will flow through the load resistance R_L. For a typical general purpose transistor, the emitter junction will have a resistance about 100 ohms.

Suppose that v_e is changed by 1 mV. The change in emitter current will be 10 μA. If we assume α for the transistor to be 0·98, the change in collector current will be 9·8 μA and this will result in a change in voltage across the 10,000 ohm load

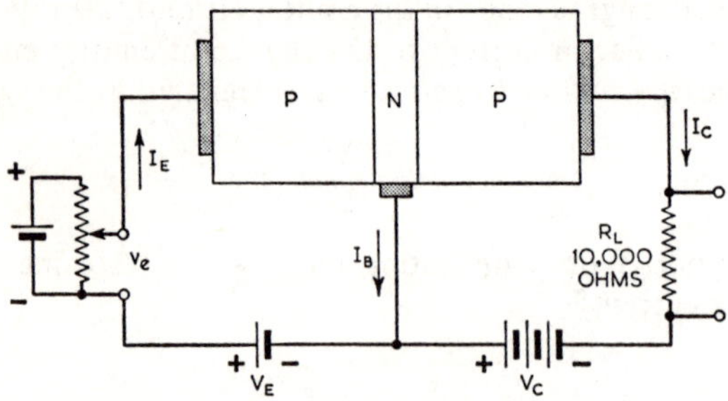

Fig. 2.14.—*The junction transistor as an amplifier.*

resistance of 98 mV. Thus although there is slight current loss we have a voltage amplification of 98 times. The amplification results from the current originally produced through the low resistance emitter junction flowing through the high resistance collector load circuit.

npn Junction transistor

An npn structure will operate in a similar way to the pnp structure. There are two main differences. First, in order to bias the emitter junction in the forward direction the n-type emitter must be made negative with respect to the base, and to bias the collector in the reverse direction the collector must be made positive. Thus the operating potentials are opposite to those for the pnp type. Secondly, the n-type emitter will inject free electrons into the p-type base and having diffused to the space charge region they will be attracted across the collector junction by the positive potential. It must be remembered that these free electrons are the minority carriers in the p-type base region.

Circuit symbols

The circuit symbols used for the pnp and npn transistors are shown in Fig. 2.15. The emitter is shown as an arrow indicating the direction of current flow (opposite to the direction of electron flow).

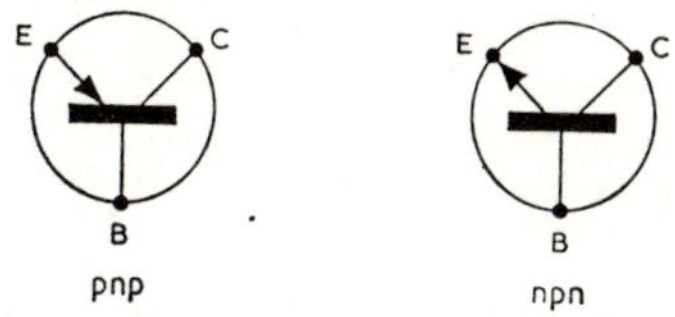

Fig. 2.15.—*Circuit symbols for transistors.*

Common base circuit

So far, we have been considering the transistor with a current fed into the emitter, the base being common to both input and output circuits. This arrangement is called the *common base circuit*, and is illustrated using the circuit symbols in Fig. 2.16.

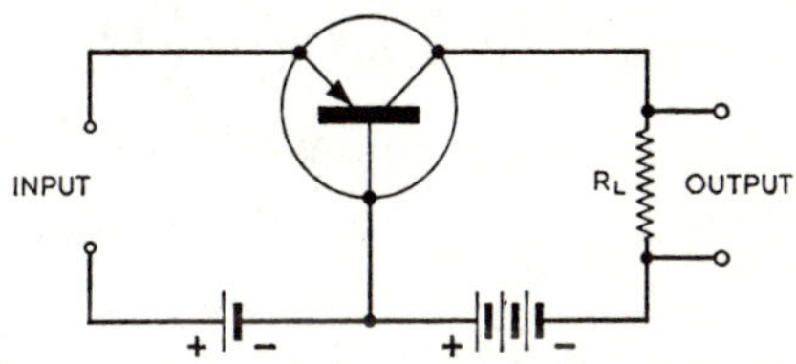

Fig. 2.16.—*The common base circuit.*

Common emitter circuit

A more convenient method of using the transistor is with the input fed into the base and the output taken from the collector,

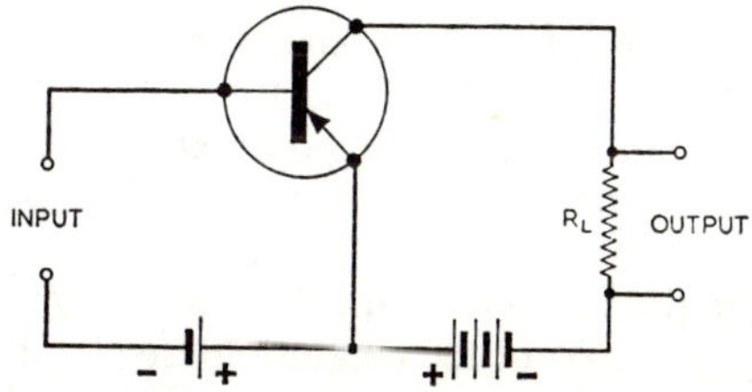

Fig. 2.17.—*The common emitter circuit.*

with the emitter common to both circuits. This is illustrated in Fig. 2.17 and is called the *common emitter circuit*. We now have a new current transfer ratio, the transfer from base to collector, which is called the *common emitter current transfer ratio* and is usually denoted by the symbol β, i.e.

$$\beta = \frac{\Delta I_C}{\Delta I_B}$$

Also, we have
$$\alpha = \frac{\Delta I_C}{\Delta I_E}$$

and
$$\Delta I_E = \Delta I_B + \Delta I_C$$

Substituting $\Delta I_C/\alpha$ for I_E and $\Delta I_C/\beta$ for I_B

we have
$$\frac{\Delta I_C}{\alpha} = \frac{\Delta I_C}{\beta} + \Delta I_C$$

$$\frac{1}{\alpha} = \frac{1}{\beta} + 1$$

Thus
$$\beta = \frac{\alpha}{(1-\alpha)}$$

Then if $\alpha = 0{\cdot}98$,
$$\beta = \frac{0{\cdot}98}{(1-0{\cdot}98)} = 49$$

Thus the common emitter circuit gives a current amplification in that a small change in the base current will result in a much larger change in the collector current. Because of this, β is usually called the *current gain*. As α, the common base current transfer ratio, nears unity, that is as the number of carriers lost by recombination in the base region becomes very small, the common emitter current amplification becomes very large.

We shall see later that the input resistance of the common emitter circuit is higher than that of the common base circuit, but is still much lower than the output resistance. Thus with the common emitter circuit both current and voltage amplification are obtained at the same time.

Common collector circuit

There is a third way in which the transistor can be arranged, that is with the signal fed into the base and the output taken from the emitter. This arrangement is called the *common collector circuit*, and is shown in Fig. 2.18. The current gain will

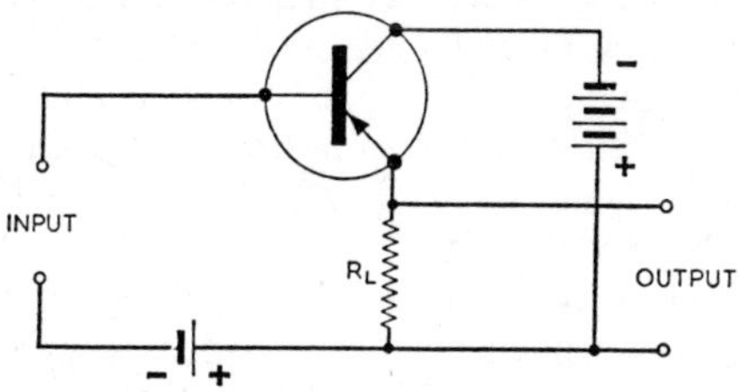

Fig. 2.18.—*The common collector circuit.*

be equal to the ratio of emitter current change to base current change.

$$\text{Current gain} = \frac{\Delta I_E}{\Delta I_B}$$

Also, we have
$$\alpha = \frac{\Delta I_C}{\Delta I_E}$$

and
$$\Delta I_E = \Delta I_B + \Delta I_C$$

Dividing by ΔI_E
$$1 = \frac{\Delta I_B}{\Delta I_E} + \frac{\Delta I_C}{\Delta I_E}$$

Thus current gain
$$\frac{\Delta I_E}{\Delta I_B} = \frac{1}{\left(1 - \frac{\Delta I_C}{\Delta I_E}\right)} = \frac{1}{(1-\alpha)}$$

It will be seen that this current gain is approximately equal to the common emitter current gain β. The output voltage, however, must always be less than the input voltage by the amount dropped between the base and emitter of the transistor.

This arrangement is often called an *emitter-follower circuit*, as the emitter voltage tends to follow the input voltage.

3
TRANSISTOR CHARACTERISTICS

It has been shown in the previous chapter that the transistor is capable of amplifying electric currents. As a result, it can be used for many applications in electronic circuits, such as audio amplifiers, hearing aids, public address amplifiers, radio and television receivers, instrumentation and industrial control. Also it will be described later how the transistor can be used as an "electronic switch", that is, it can present either a high or a low resistance to the passage of current. This opens up the possibility of wide use in computer circuits and control systems.

For each application careful circuit design work must be carried out. Before this can be done systematically, it is necessary to have detailed knowledge of the characteristics of the transistor as a circuit element, that is, to know what is the best operating voltage and current, what impedance is presented to the signal, what amplification the transistor will give, what is the internal impedance of the transistor at the output, and so on. Data from which information of this nature can be obtained is prepared by the manufacturer on each type of transistor and published as "data sheets" so that the user can carry out his preliminary design without having to make measurements himself.

First, it is important to derive groups of voltage–current relationships for the transistor. Two sets of curves are normally presented, the forward voltage–current characteristics of the emitter junction, referred to as the *emitter characteristics* or *input characteristics*, and the reverse voltage–current curves of the collector junction, called the *collector characteristics* or the *output characteristics*.

Common base characteristics

Considering first the common base arrangement, Fig. 3.1 shows the emitter to base forward characteristic, that is, how the emitter current increases as the emitter to base voltage is increased positively from zero. The characteristic shown is typical of a small signal germanium transistor. It will be seen that at first the current increases only very slightly as the

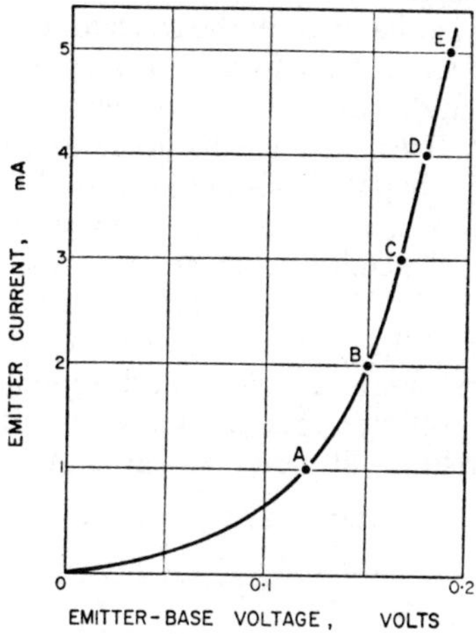

Fig. 3.1.—*Input characteristic of common base circuit.*

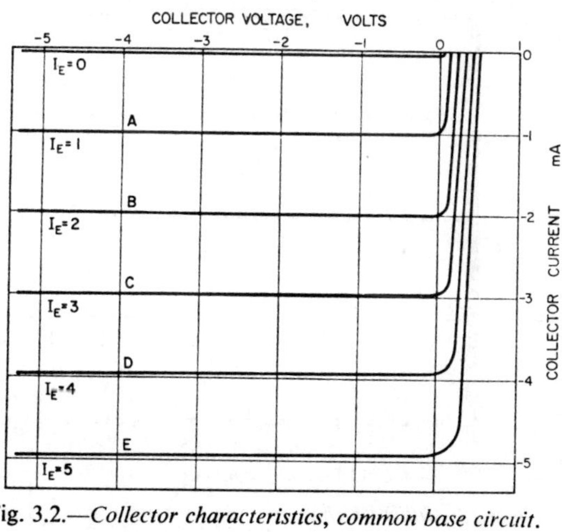

Fig. 3.2.—*Collector characteristics, common base circuit.*

voltage is increased. During this region the applied voltage is overcoming the potential barrier of the junction (as described in Chapter 2). Once the barrier has been neutralised, the current increases rapidly.

Now consider what happens in the collector circuit when the emitter current is varied. At first, with zero emitter current, the collector to base characteristic is shown as the curve for $I_E = 0$ in Fig. 3.2. This is similar to the reverse characteristic of a pn junction shown previously in Fig. 2.10, (Chapter 2). The small current is known as the *leakage current* of the collector junction. Now let the emitter current be increased to 1 mA—point A in Fig. 3.1—and held constant at that value. We have seen that nearly all of the emitter current passes to the collector, the amount of current crossing the collector junction not being dependent on the collector voltage. Thus the collector voltage current characteristic will now be curve A for $I_E = 1$ in

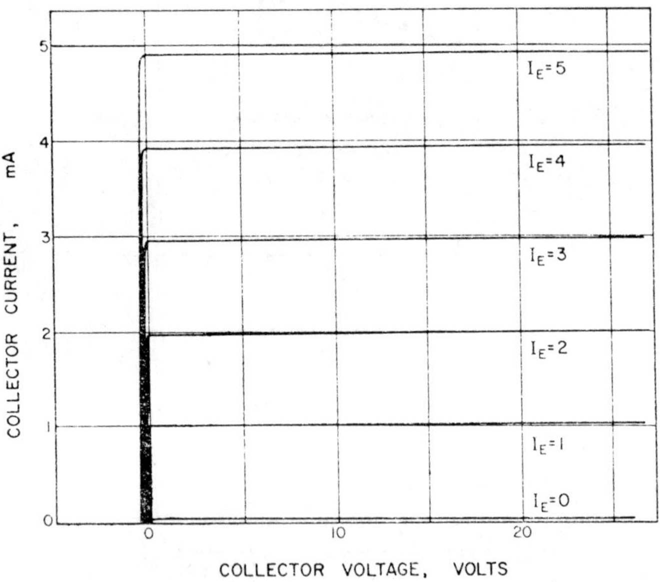

Fig. 3.3.—*Collector characteristics, common base circuit.*

Fig. 3.2. Similarly, as the emitter current is increased in further steps, collector curves B, C, D and E are obtained for emitter current $I_E = 2, 3, 4$ and 5 mA. The almost horizontal nature of the collector characteristics emphasises the high output resistance, a large change of collector voltage producing only a very small change of current.

It will be seen that the collector current is maintained even at zero collector voltage. This is because the base current, in flowing out through the resistance of the base region, sets up a small potential which appears in the collector–base circuit, and constitutes a small reverse bias across the collector junction. To reduce the collector current to zero it is necessary to apply a small forward collector voltage as shown in Fig. 3.2.

In Fig. 3.2 the collector characteristics have been shown in the third quadrant as a reminder that the collector junction is biased in the reverse direction. It is now customary to present them in the first quadrant as shown in Fig. 3.3, to some extent because the output characteristics of thermionic valves were always drawn in this way.

Common emitter characteristics

With the common emitter connection, similar characteristics can be prepared. First the input characteristic, which shows how the base current varies as the voltage across the emitter–base junction is increased in the forward direction, as shown in Fig. 3.4, and secondly, the output characteristics, showing how the collector current varies with collector to emitter voltage for different levels of base current, as shown in Fig. 3.5. It will be seen from Fig. 3.4 that the input resistance is higher than for the common base arrangement, the change of base current being smaller than the change of emitter current for a given change of emitter to base voltage. The output characteristics in Fig. 3.5 are similar to the common base characteristics except that there is now a noticeable slope on the current lines, the current increasing with voltage. This indicates that the output resistance is lower than in the common base arrangement; but nevertheless it is still high. Also, the collector current is now zero for zero collector voltage since the potential

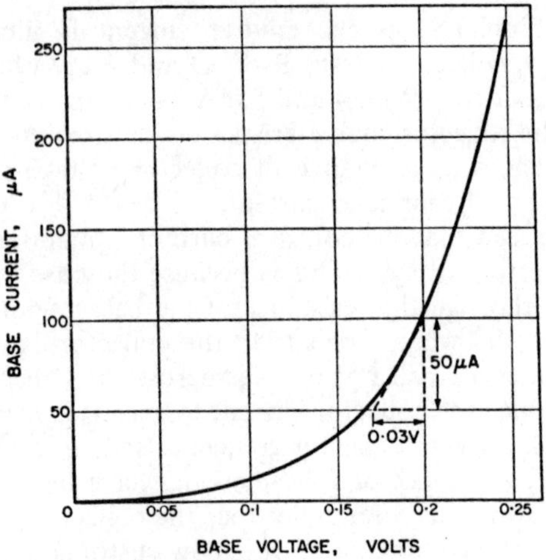

Fig. 3.4.—*Input characteristic of common emitter circuit.*

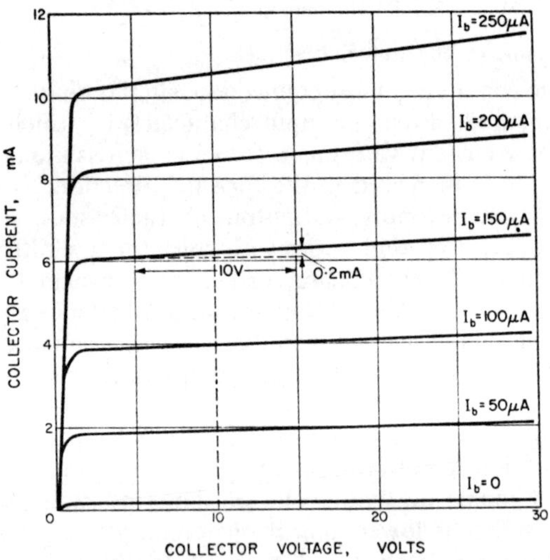

Fig. 3.5.—*Collector characteristics, common emitter circuit.*

produced by the base current does not appear in the collector to emitter circuit.

D.C. operating point

These characteristic curves are immediately useful in helping to choose a suitable d.c. operating point for amplifier applications. This topic will be discussed in more detail later.

Input and output resistances

The a.c. input and output resistances of a transistor can be determined from the slopes of the input and output characteristics at the operating point. Referring to Fig. 3.4, if we consider a transistor operating at 100 μA base current, the slope of the characteristic at that point is $0{\cdot}03/(50 \times 10^{-6})$ giving a slope resistance of 600 ohms. This is the input resistance of the transistor in the common emitter connection. In Fig. 3.5, at 6 mA collector current, the slope of the collector characteristic is $10/(0{\cdot}2 \times 10^{-3})$ giving a slope resistance of 50,000 ohms. This is the output resistance of the transistor in the common emitter connection.

Current transfer characteristics

It is possible to deduce the current transfer—or current gain—characteristics from the family of collector characteristic curves. Considering the common emitter characteristics of Fig. 3.5, suppose it is required to determine the current gain at a collector current of 5 mA and a collector voltage of 10 volts. If a line is drawn vertically at the collector voltage of 10 volts, the change of collector current for a change of base current from 100 μA to 150 μA is 4 mA to 6·1 mA. The current gain is the change of collector current divided by the change of base current, i.e. in this case 2·1 mA divided by 50 μA, which gives a value of 42.

The transfer characteristic is, however, usually presented in a separate curve. It was customary to give a curve of collector current versus base current for a constant collector voltage, as

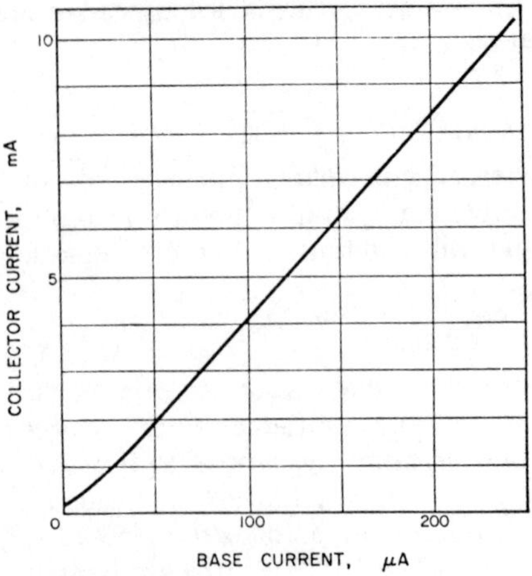

Fig. 3.6.—*Current transfer characteristic.*

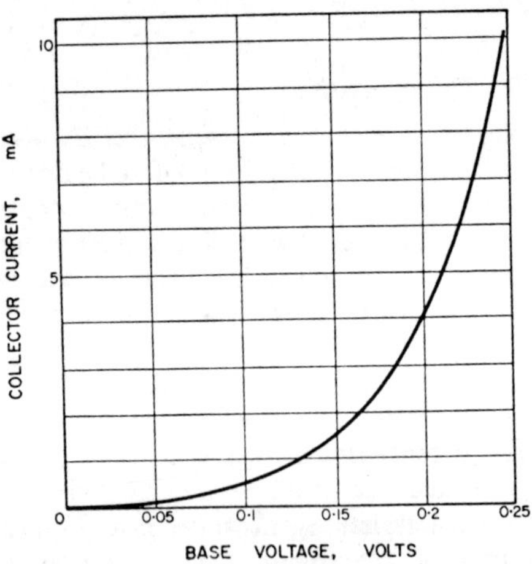

Fig. 3.7.—*Hybrid transfer characteristic.*

shown in Fig. 3.6, but now it is more usual to give a curve of collector current versus base voltage as in Fig. 3.7. This latter curve is more useful in switching applications where emitter voltage is varied to switch the collector current.

Collector leakage current

The collector leakage current is the small saturation current flowing across the reverse biased collector junction. In the common base connection, this leakage current is given the letter symbols I_{CBO} (the leakage current *I* from *C*ollector to *B*ase with emitter *O*pen circuit), and in the pnp transistor flows out of the base as a negative base current $(-I_{CBO})$.

In the common emitter connection with the base open circuit this leakage current must flow out through the emitter junction. Since it would normally flow out through the base, in order to make base current zero, we must imagine a current equal to $+I_{CBO}$ fed into the base to cancel it out. This current of $+I_{CBO}$ will be amplified by the transistor and a current of βI_{CBO} will appear in the collector circuit. This common emitter leakage current is called I_{CEO} (leakage current *I* from *C*ollector to *E*mitter with base *O*pen circuit).

The leakage current is a very important parameter, as it increases exponentially with temperature and can set a limit to the operating temperature as discussed in the next paragraph.

Effects of temperature

With germanium transistors, the leakage current roughly doubles for every 9°C rise in temperature, and with silicon transistors for every 11°C. This is illustrated in Fig. 3.8 which shows the variation of the leakage currents of typical small signal germanium and silicon transistors. It will be seen that with the germanium transistor, which may have a common emitter leakage current about 50 μA at room temperature (20°C), the current will have increased to the order of 2 mA at 75°C. Silicon transistors have a much lower leakage current at room temperature, and it will not have increased to the same order until the temperature is about 200°C.

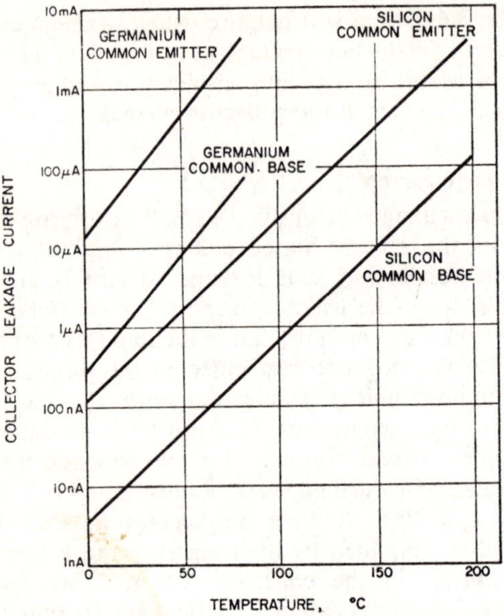

Fig. 3.8.—*Variation of collector leakage current with temperature.*

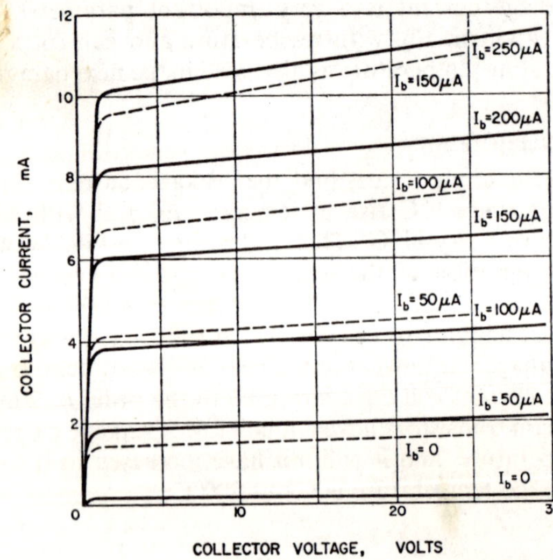

Fig. 3.9.—*Effect of temperature on collector characteristics.*

Increasing the temperature has a marked effect on the collector characteristics. This is illustrated in Fig. 3.9. First, the curve for $I_b = 0$, the leakage current, will be much higher, as shown by the bottom dotted line. Increasing the base current will increase the collector current above the new $I_b = 0$ curve, that is, the whole family of curves will be higher. Secondly, the change in collector current for a given change in base current is higher than for room temperature, that is, the current gain is higher. The fact that the collector current increases so much with temperature means that an operating current chosen at room temperature will also increase, and care must be taken to ensure that the additional power loss can be safely dissipated in the transistor.

Ratings and operating limitations of transistors

In addition to describing the properties and characteristics of transistors, it is necessary to state over what range of operating conditions it is safe to operate them, for instance the maximum voltages and power dissipation that can be used, the maximum ambient and internal temperatures, and so on. Also there are other limitations to the operation of a transistor which depend upon its design and fabrication, such as the highest frequency at which it will operate and the level of the self-generated noise.

Maximum collector voltage

It was stated in Chapter 2 that if the reverse voltage across a pn junction exceeds a certain value breakdown occurs, a high reverse current flows, and the resistance of the junction falls to a low value. These considerations apply to the collector junction of a transistor, and if the junction breaks down in this way, the transistor will no longer function and, in fact, it may be destroyed if suitable precautions are not taken to limit the current. The breakdown voltage of the collector junction is known as BV_{CBO}—the *B*reakdown *V*oltage, *C*ollector to *B*ase with emitter *O*pen circuit. It is measured by applying a reverse voltage between collector and base, and increasing it until a certain predetermined current is reached. The manufacturer usually allows a factor of safety and quotes a value somewhat lower than the measured value.

The collector voltage can also be limited by another effect. The reverse voltage creates a depletion layer at the junction. This depletion layer becomes wider as the voltage is increased and eventually may extend completely across the base region to the emitter. Beyond this voltage, the transistor would appear as a straight resistance from emitter to collector and transistor action would be lost. This effect is known as *punch-through*, and can happen at a voltage lower than the collector junction breakdown voltage, although this is not usual in correctly designed transistors.

Maximum emitter reverse voltage

The emitter junction is normally operated with forward bias, but in certain applications it is necessary to reverse bias the junction. For example, in switching circuits, when it is desired to "cut-off" the collector current, the emitter junction is reverse biased. Since the emitter region has a high carrier concentration in order to obtain a high injection efficiency, the reverse breakdown voltage of the emitter junction is usually lower than that of the collector junction.

Collector current limitation

High collector current in itself is rarely a limitation in transistors, the current carrying capacity being usually much higher than the limiting value which is determined either by maximum power dissipation or by a loss of performance of the transistor.

Maximum junction temperature and power dissipation

It has already been described how the collector leakage current increases rapidly with temperature. Eventually it will be so high as to limit the usefulness of the transistor. With germanium the limiting temperature is of the order of 85°C and with silicon about 200°C. This limit refers to the temperature of the collector junction. When a transistor is operating, power is dissipated due to to the collector voltage and current.

This power, equal to the product of the voltage and current, results in the generation of heat, and causes a rise in temperature of the junction. The increase in temperature will depend on how fast the heat can flow away, and the temperature will

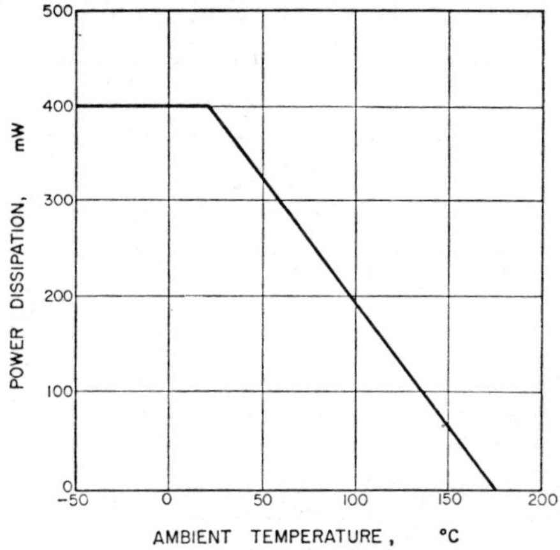

Fig. 3.10.—*Power derating characteristic of a silicon transistor.*

continue to rise until the rate of heat generated by the power dissipation equals the rate of heat flowing away. The property of a transistor to loose heat is expressed as its *thermal resistance*, and is quoted by manufacturers as the junction temperature rise per unit power dissipated. From this figure the user can determine the maximum allowable power dissipation for any ambient temperature, by first determining the allowable temperature rise and dividing it by the thermal resistance. For example a silicon transistor may have a thermal resistance from junction to ambient room temperature of 0·375 mW per degree centigrade and a maximum junction temperature of 175°C. With the ambient temperature at 20°C the maximum allowable power dissipation is $155°C \times 1/0{\cdot}375 = 400$ mW. With an ambient temperature of 100°C, the power dissipation will be limited to $75°C \times 1/0{\cdot}375 \simeq 200$ mW. The manufacturer often gives a power derating curve as shown in Fig. 3.10.

Frequency limitation

If the frequency of a signal applied to a transistor is increased, the output initially remains constant, but eventually it will fall off due to the amplification of the transistor decreasing. There are two main causes for this. First, the

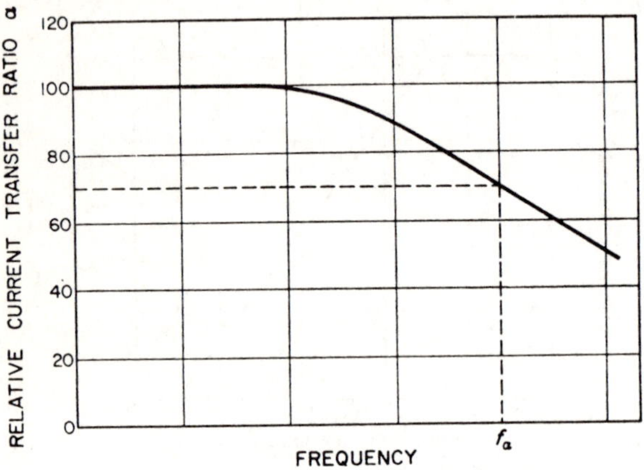

Fig. 3.11.—*Frequency/gain characteristic.*

minority carriers take a finite time to diffuse across the base region from the emitter junction to the collector junction. This time is called the *transit time* and increases as the thickness of the base region increases. Thus, for a transistor to have a good performance at high frequency the base width must be very small. To illustrate this, a pnp germanium alloy transistor will typically have a base width of 0·001 inch and a cut-off frequency of 3 Mc/s. An npn silicon diffused transistor might have a base width of 0·0001 inch with a cut-off frequency of 150 Mc/s. The "cut-off frequency" is the frequency at which the common base current transfer ratio has fallen to 0·7 of its initial value and is called f_α. The general form of the frequency characteristic is shown in Fig. 3.11.

The second limitation is the capacitance of the collector junction. This effectively shunts the junction at high frequency. Also, in conjunction with the series resistance of the base region, it can cause negative feedback, which increases with frequency. These effects will be discussed in more detail in later chapters.

Noise characteristics

The usefulness of a transistor in amplifying very small signals is eventually limited by "noise" generated in the

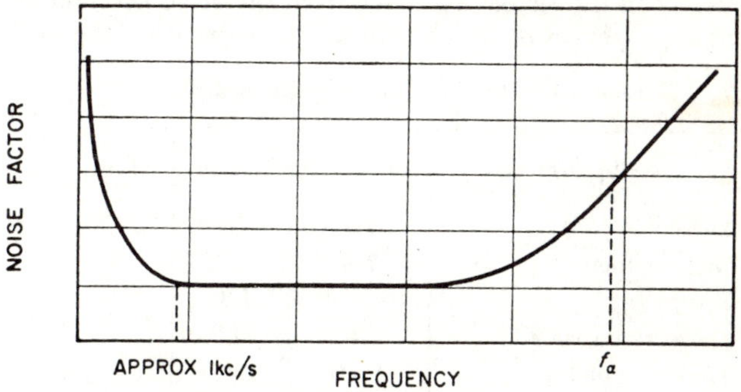

Fig. 3.12.—*Form of noise characteristic.*

transistor itself. At low frequencies, below about 1 kc/s, the noise level increases as the frequency decreases. Above 1 kc/s the noise level remains approximately constant up to the cut-off frequency. Above this frequency the noise level increases as the amplification decreases. The general form of the noise characteristic is shown in Fig. 3.12.

4

TRANSISTOR TECHNOLOGY

THE first step in the manufacture of transistors is the processing of the semiconductor material into the required form. Practically all transistors are made from silicon or germanium, the main properties of which are listed in Table 4.1. The main

Table 4.1. Properties of germanium and silicon

Property	*Germanium*	*Silicon*	*Unit*
Density	5·3	2·3	gm/cc
Melting point	950	1,425	°C
Energy gap	0·72	1·08	eV
Dielectric constant	16	12	—
Electron mobility	4,000	1,400	cm/sec per volt/cm
Hole mobility	1,900	500	cm/sec per volt/cm
Intrinsic resistivity (25°C)	65	200,000	ohms

difference between these two materials from the transistor viewpoint is the energy gap, which is a measure of the energy that has to be given to an electron before it will break away from its atom. For germanium, the energy gap is 0·72 eV and for silicon 1·08 eV. The fact that the energy gap is higher for silicon means that with silicon it is necessary to increase the energy of an electron to a higher level in order to break it free from its parent atom. It would, for example, be necessary to raise the silicon to a higher temperature for electrons to break free. Thus at any given temperature less electrons and holes will be thermally generated in silicon than in germanium.

The reverse leakage current of a pn junction consists of holes and electrons thermally generated around the junction, and so at room temperature the leakage current of a silicon pn junction will be much less than for a germanium junction. This means that a silicon pn junction can be raised to a higher temperature before its leakage current is equal to that of a germanium junction.

In a transistor these considerations of reverse leakage current apply to the collector junction and so silicon transistors can operate with the collector junction at relatively high temperatures—up to 200°C, compared with germanium transistors which are limited to about 85°C. Thus a higher internal temperature rise is allowable in a silicon transistor—and this in turn means that a higher internal power dissipation can be allowed. Silicon transistors can operate at two to three times the allowable power dissipation of similar germanium transistors.

The higher energy gap of silicon, however, also brings some disadvantages. A higher forward voltage must be applied to a silicon pn junction to overcome the potential barrier, and so the forward voltage drop for a given current will be higher than for germanium. Also silicon may be somewhat more difficult to process in view of its higher melting point—1,425°C against 950°C for germanium.

In addition to silicon and germanium, other semiconductor materials called compound semiconductors are being investigated. These compounds are formed from a combination of an element having three valence electrons with one having five valence electrons such as gallium arsenide, indium antimonide gallium antimonide, gallium phosphide and so on. Of these, gallium arsenide, which has an energy gap of 1·3 eV, may well become the most important.

Germanium

Germanium occurs in nature as a low level impurity in zinc and copper ores. Also, in the United Kingdom, it appears in certain coal formations and is extracted from the flue dust of power stations using coal. The germanium is extracted chemically, converted to germanium dioxide—a white powder—and subsequently reduced to the metal by heating in hydrogen.

At this stage, the metal could be called chemically pure—the impurities may be down to the order of one part per million. However, this is not sufficiently pure for transistor manufacture and so it is further purified by a process known as *zone refining*. Power is concentrated into one part of a long ingot of germanium so that a narrow molten zone is produced. This zone is then moved along the bar from one end to the other. Most impurities are more soluble in molten than in solid germanium, and so the impurities will prefer to stay in the molten zone as it moves along to the end of the ingot, and to settle out at the end when the zone is solidified. A succession of molten zones passing along the ingot will leave most of it very pure, with all the impurities concentrated at the end, which may be cut off and rejected. By this means, the impurities can be reduced to the order of one part in 10^{10}.

For pn junctions to have the characteristics described earlier, it is now necessary to convert the material into single crystal form, that is, to have a continuous crystal structure, and at the same time to add the required amount of the selected donor or acceptor material. For transistors the amount of donor or acceptor is about one part in a hundred million. To produce single crystal material, the germanium is placed in a crucible, heated to a few degrees above its melting point, and held constant at that temperature. A small piece of single crystal germanium, known as a "seed", is lowered into contact with the surface of the molten germanium, and then slowly raised at a rate of one or two inches per hour. As the seed is raised, the germanium solidifies on to it with a continuous crystal structure.

Silicon

Silicon is undoubtedly the most important semiconductor material at the present time. It is probably the most common of all the elements, appearing as sand, which is mainly silicon dioxide, and in various silicate compounds. As with germanium, the metal is extracted chemically. The most troublesome impurity is boron—an acceptor—and unfortunately it is not possible to remove this by the zone refining technique as described above for germanium, since it has roughly equal solubilities in molten and solid silicon. As a result, it is necessary to carry the chemical purification further, and it is now possible to reduce the boron content down to a few parts in 10^{10} by this means.

Single crystal silicon can be produced by "pulling from the melt" as described for germanium. Also it can be produced by a method known as the *floating zone process*, in which a rod of polycrystalline silicon is held vertically and a molten zone, narrow enough to be held in position by surface tension, is passed up, from a single crystal seed at the bottom, to the top. As the zone passes up, silicon melts into it at the top and solidifies out into single crystal form at the bottom. At the same time, impurities other than boron are taken up with the zone and deposited at the top. This floating zone process makes it possible to prepare single crystal material without the possibility of contamination from a crucible.

Crystal slicing and wafer preparation

To prepare the single crystal material for device fabrication it must be cut up into suitable pieces. First it is sawn into slices about fifteen-thousandths of an inch thick. For some devices—particularly those prepared by diffusion—the subsequent pn junction processes are carried out on the whole slice, but for other devices it is necessary to cut the slices into rectangular or circular "wafers" which are processed into devices individually. Slicing is usually carried out by sawing with diamond impregnated wheels. The slices are then lapped to a thickness about eight to ten thousandths of an inch using carborundum powder and finally the remaining surface damage is removed by etching in an acid mixture—typically one part concentrated hydrofluoric acid to four parts concentrated nitric acid. When it is required to cut the slice into wafers this is carried out on the lapped slices by scribing (similar to glass cutting), ultrasonic cutting, by sawing with a diamond impregnated wheel or sawing with an abrasive loaded tungsten wire.

pn Junction formation

It has been described earlier how a pn junction is a transition from p-type to n-type material. It is important to remember that the transition must occur in a continuous single crystal lattice in which the semiconductor atoms in the p-type region are oriented in exactly the same way as those in the n-type region. It is impossible to obtain a pn junction by making separate p-type and n-type pieces of semiconductor and bringing them together into contact.

The first junction transistors had pn junctions formed during the single crystal growth by pulling from the melt. These were called *grown junction transistors* and were formed as follows. The growth of an n-type crystal was started, and after it had grown about a quarter of an inch long a pellet of acceptor material was dropped into the molten silicon so as to convert the growing crystal from n-type to p-type. Soon after, a second pellet, this time of a donor material, was dropped into the melt and the growing crystal changed back to n-type. This resulted in an npn crystal with a narrow p-type layer. The crystal was cut up into small bars with the p-type layer central. Each bar was mounted up to form an npn transistor by making ohmic contacts to the two ends and to the p-type layer. This type of transistor, however, is now rarely manufactured and so no further mention of it will be made.

Other than this grown junction method, there are three basic processes for producing pn junctions—alloying, diffusion and epitaxy. These processes are all of great importance and will be described in some detail.

Alloy junction

To describe the formation of an alloy junction, it is convenient to consider the particular case of alloying indium with germanium. The melting point of indium is 156°C and of germanium 950°C. The steps in the process will be described with reference to Fig. 4.1. A small sphere of indium (an acceptor) is positioned on an n-type germanium wafer and the combination heated. At 156°C the indium melts and becomes a molten "droplet". As the temperature increases further the molten indium starts to dissolve some of the germanium, the amount dissolved increasing as the temperature rises. As more germanium goes into solution, the boundary between the solid germanium and the molten indium–germanium solution,

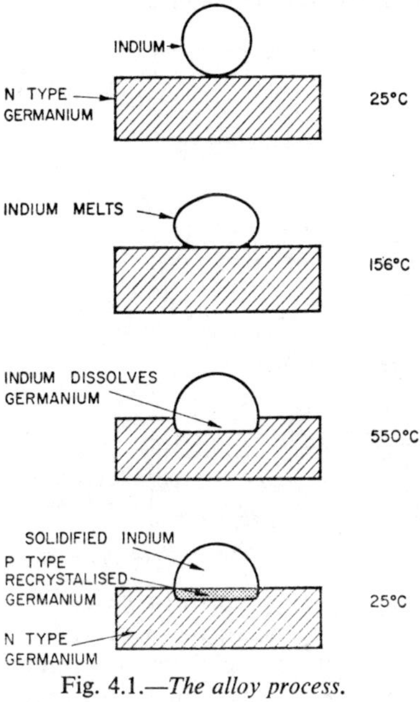

Fig. 4.1.—*The alloy process.*

known as the *solid–liquid interface*, moves into the germanium wafer. The temperature is increased to the order of 550–600°C and then reduced to room temperature. As the combination cools, the germanium recrystallises out of solution back on to the wafer, with the same crystal orientation, but now it is heavily doped with indium, and so the recrystallised layer is p type thus forming a pn junction with the original n-type wafer.

In a similar way a silicon alloy junction can be formed by alloying aluminium (another acceptor) to an n-type silicon wafer.

Diffused and planar junctions

As opposed to the alloy process, which involves a liquid phase, diffused junctions are made by the movement of acceptor or donor atoms into the solid semiconductor wafer. If an n-type semiconductor wafer is heated to a temperature

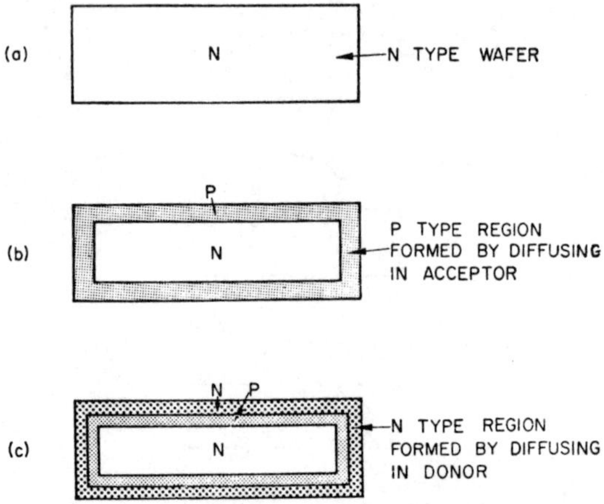

Fig. 4.2.—*The diffusion process.*

approaching its melting point in a gaseous atmosphere containing an acceptor, some of the acceptor atoms will diffuse into the semiconductor and change the original n-type surface layer over to p type by introducing an excess of acceptors. Thus a pn junction is formed a small distance inside the wafer as shown in Fig. 4.2(b). In a similar manner, diffusion will also take place from a solid donor or acceptor element which is in contact with the surface of the semiconductor. With silicon, a temperature of about 1,000–1,200°C is usual and with germanium about 850°C. Even at these high temperatures the rate of diffusion into the semiconductor is very slow, a typical rate being about one ten-thousandth of an inch per hour.

Convenient diffusants with silicon are boron as an acceptor and phosphorus as a donor, and with germanium, antimony as a donor and gallium as an acceptor.

It is possible to take an n-type semiconductor wafer, diffuse in an acceptor to give a p-type layer, and then to diffuse in a donor to convert the surface back to n type and result in an npn structure as illustrated in Fig. 4.2 (c).

An important fact is that silicon oxide prevents the diffusion of some elements and this can be used to control the location of diffused regions in the so-called "planar process", as will be shown later.

Epitaxial junctions

In epitaxial crystal growth, a thin film of semiconductor is grown on to a single crystal wafer of the same material, the crystal orientation being continued from the wafer into the layer—the word *epitaxial* means "same axis". The wafer on which the epitaxial layer is grown is called the *substrate.*

For the epitaxial growth of silicon, a volatile silicon compound such as silicon tetrachloride is mixed with hydrogen and the mixture fed into a chamber where the silicon substrate slice is located. The slice is heated to a high temperature, about 1,100°C, to promote a chemical reaction with the vapour. The silicon tetrachloride dissociates, the hydrogen forms a compound with the chlorine, and silicon is deposited on to the substrate slice to form the epitaxial layer. A typical rate of growth of the epitaxial layer is one micron (0·00004 inch) per minute. The conductivity of the epitaxial layer can be arranged to be either p type or n type by introducing an acceptor or donor on to the hydrogen stream with the silicon tetrachloride vapour.

It will be seen that by starting with a p-type substrate slice and growing an n-type epitaxial layer, a pn junction can be formed. Alternatively a high resistivity n-type layer can be grown on a low resistivity n-type substrate—this is useful in certain devices, as will be discussed later.

Transistor fabrication

A transistor consists essentially of three different regions in a piece of semiconductor material separated by two pn junctions. Thus in fabrication it is necessary to form the two

Table 4.2. Typical

Material	Type	Dimensions		
		Emitter (in.)	*Collector* (in.)	*Base width* (Thousandths of an inch)
Germanium	Alloy	0·011 dia.	0·017 dia.	0·3
Germanium	Alloy power	0·12 dia.	0·18 dia.	0·7
Germanium	Alloy diffused	0·005 dia.	0·006 × 0·012	0·1
Germanium	Diffused mesa	0·003 × 0·001	0·004 square	0·03
Silicon	Alloy	0·033 dia.	0·044 dia.	0·5
Silicon	Diffused mesa small signal	0·006 × 0·002	0·008 square	0·05
Silicon	Epitaxial planar small signal	0·005 dia.	0·010 dia.	0·05
Silicon	Epitaxial planar medium power	0·015 dia.	0·032 dia.	0·08
Silicon	Epitaxial planar high power	"E" shape 0·15 × 0·1	0·2 square	0·08

pn junctions in the required locations, and then make ohmic connections to the three regions. The optimum requirements of the two pn junctions may be different for certain types of transistor, and so it may be desirable to use different methods of junction formation. This, in fact, is done in the alloy-diffused and diffused-base transistors. In most cases, however, the emitter and collector junctions are both formed by the same process as in the alloy, diffused mesa, and planar transistors.

Table 4.2 gives tabulated data on the various categories of transistor available.

Germanium alloy transistor

The basic structure of a germanium alloy transistor is shown in Fig. 4.3. Two pellets of indium are alloyed to the opposite faces of a thin n-type germanium wafer to form the emitter and collector junctions. Also, a metal tab is fused to the wafer to form the ohmic contact to the base region. During the alloying operation, the molten indium on each side moves

transistor characteristics

Cut-off frequency f_T (Mc/s)	*Maximum ratings*				Collector leakage current I_{CBO} (μA) 20°C	Collector base capacitance at 6V (pF)
	Collector voltage (V)	*Collector current* (A)	*Collector power* (W)	*Junction temperature* (°C)		
10	25	0·2	0·2	85	1	20
0·4	60	7	150 Watts case at 25°C	100	200	200
150	40	0·03	0·1	85	1	12
800	30	0·05	0·1	85	1	2
1	40	0·1	0·3	200	0·01	40
250	60	0·1	0·3	200	0·01	8
300	80	0·1	0·3	200	0·001	8
50	80	0·5	0·8	200	0·01	20
10	80	7	50 Watts case at 100°C	200	0·1	500

into the germanium wafer. The indium thickness, alloying temperature and wafer thickness are all precisely controlled so that the final spacing between the two junctions is between

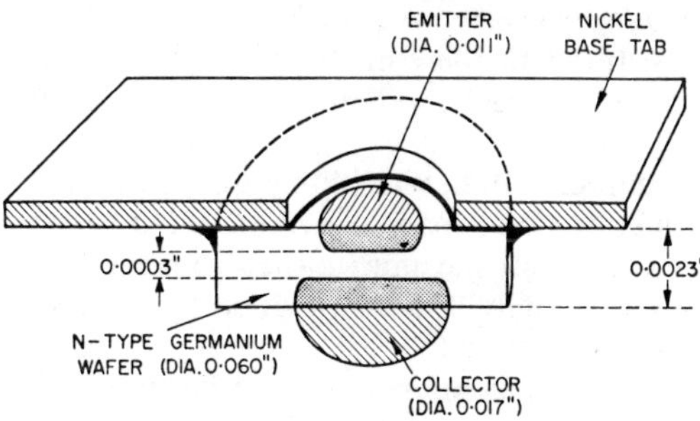

Fig. 4.3.—*Germanium alloy transistor element.*

a half and one thousandth of an inch. In practice the indium and germanium wafer dimensions are fixed in the design specification and the required base-width obtained on the production line by fine adjustment of the alloying temperature, which is in the region of 600°C. Typical dimensions of a conventional low power germanium alloy transistor are indicated in Fig. 4.3. The collector junction is larger than the emitter junction to ensure that as many as possible of the holes, which spread out from the emitter, are collected to result in a high current gain.

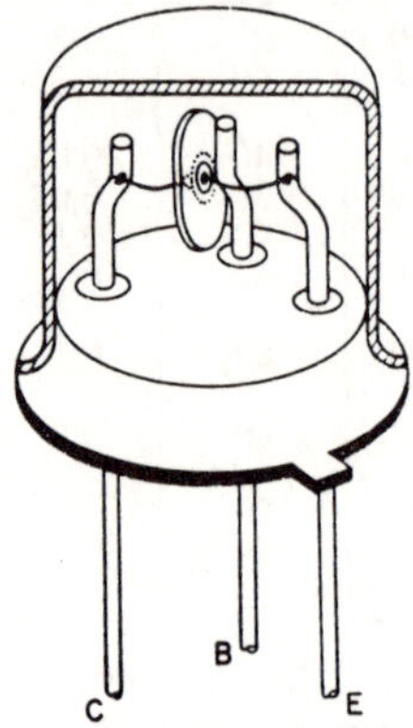

Fig. 4.4.—*Cut-away view of a germanium alloy transistor.*

The basic unit described above is mounted to a header, consisting of a platform with three insulated through leads. Small wires, fused to the emitter and collector indium dots, are welded to two leads, and the base tab is welded to the third.

Surface contamination, which may bridge the junctions and result in inferior performance, is removed by electrolytic or chemical etching and the unit is then thoroughly washed in purified water. Finally, the unit is dried by prolonged baking at 110°C and encapsulated by welding a metal can to the header. Some manufacturers cover the junctions with varnish or silicone grease in an attempt to give further surface protection, or in some cases to give better heat conduction from the transistor unit to the can. A cross section of a completed unit is shown in Fig. 4.4.

A typical low power germanium transistor may have a maximum collector voltage between 15 and 50 volts, a maximum power dissipation of 200 mW a current gain of 50 and a frequency cut-off between 2 and 12 Mc/s.

Important advantages of the alloy transistor are:

(1) The contacts to the collector and emitter regions are mainly metallic indium, and so the series resistance between the junctions and outside leads is low. This results in a low value of voltage dropped across the transistor in the saturated condition.

(2) The emitter can withstand a high reverse voltage—of the same order as the collector.

(3) Since there are a small number of steps in the manufacturing process, the alloy transistor is basically a low cost device.

The main disadvantages are:

(1) It is difficult to control the formation of very thin base layers by alloying in from each side of a wafer, and so the process is not suitable for the production of high frequency transistors.

(2) The resistivity of the base region is higher than that of the collector recrystallised layer, and so the collector depletion layer will move mainly into the base region. This sets an upper limit to the collector voltage.

Germanium alloy power transistor

The basic alloy process can be extended to larger area junctions for low frequency power transistors. Typical emitter and collector diameters for a power transistor capable of operation at several amperes collector current are 0·070 and 0·10 inch respectively. The base width must still be kept as low as possible—of the order of 0·001 inch, to obtain a high current gain characteristic. An essential feature of the construction of a power transistor is that a low thermal resistance heat path must be provided so that heat generated in the transistor can readily flow out to an external heat radiating system. A typical construction for a germanium alloy power transistor is shown in Fig. 4.5.

The allowable power dissipation in the transistor is dependent upon the ambient temperature, the internal thermal resistance between collector junction and case, and the thermal resistance of the external heat radiating system to air. The whole question is somewhat complicated to specify, but manufacturers provide extensive details in their data sheets, and these must be meticulously followed.

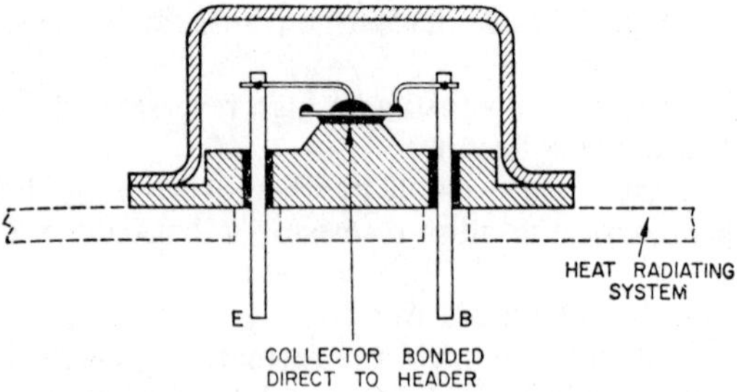

Fig. 4.5.—*Cross-section of a germanium alloy power transistor.*

Silicon alloy transistor

Silicon alloy transistors are made by alloying aluminium into n-type silicon. The general considerations are similar to those for the germanium alloy transistor. A different construction is necessary, see Fig. 4.6, to allow for the fact that silicon–aluminium alloy is quite brittle and has a different expansion coefficient from silicon. Thus a strain can be set up in the silicon on solidification after alloying. Indium is a very soft metal and so does not set up any strain in the germanium alloy transistor. For the silicon device, thin discs of aluminium foil are used to alloy to the silicon, and are backed by molybdenum electrodes, which have an expansion coefficient similar to that of silicon. The alloying temperature is of the order of 600°C and, as in the germanium alloy transistor, is adjusted to give the value of base width required. The resulting unit is mounted, etched, dried and encapsulated in a manner basically similar to that for the germanium alloy transistor. Since the silicon transistor can be operated up to 200°C junction temperature it is normal during drying to bake the unit at a temperature somewhat higher than this, about 220°C, to ensure that the final traces of water vapour are driven off and the surface has stabilised before encapsulation.

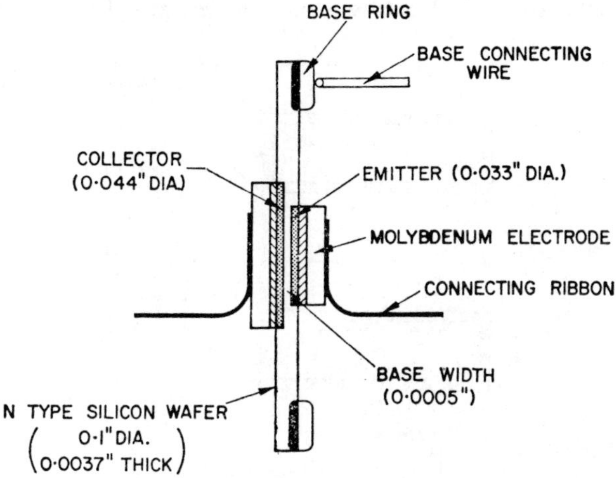

Fig. 4.6.—*Silicon alloy transistor element.*

The main advantages of the silicon alloy transistor are:

(1) The low collector leakage current, which is typically of the order of 10 nanoamperes (10^{-8} amps).

(2) The high operating temperature—up to 200°C junction temperature—which allows operation in ambient temperatures up to the order of 175°C or alternatively allows a power dissipation of about 200 mW at an ambient temperature of 30°C.

(3) As with germanium, it is basically a low cost device.

Its main disadvantage is its low frequency cut-off which is lower than the germanium alloy transistor due to the lower mobility of silicon.

Alloy diffused transistor

The alloy diffused transistor uses a combination of the alloying and diffusion processes to give a narrower base width and so a better high frequency performance. The basic steps in the process are illustrated in Fig. 4.7. A p-type germanium wafer is taken and a thin n-type layer formed around the surface by diffusion. This layer will become the base layer of the finished transistor. Next, two small pellets are alloyed to the

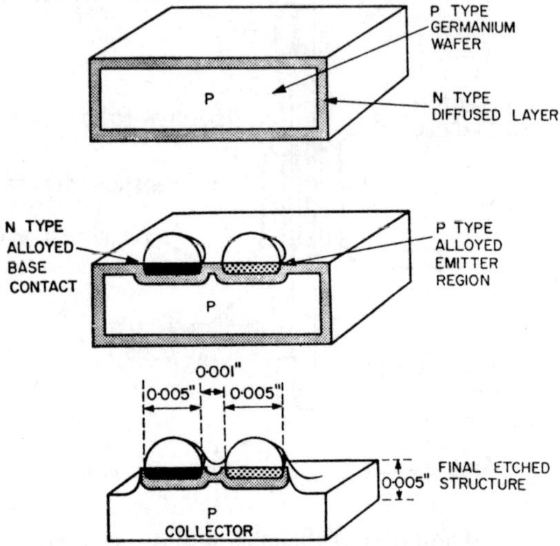

Fig. 4.7.—*Basic steps in the alloy diffused process.*

top of the wafer side by side. The first pellet contains an n-type material and so makes ohmic contact to the n-type diffused layer and becomes the base contact. The second pellet contains mainly p-type material and on alloying to the n-type diffused layer forms the emitter junction; but it also contains some n-type material. Since both pellets contain some n-type material the diffused n-type region is pushed further into the wafer under the alloying pellets. The area between and around the two pellets is then masked, and the n-type layer elsewhere is removed by etching. This defines the area of the collector junction. Base width can be controlled down to about 0·0005 inch giving frequency cut-off up to 100 Mc/s, and high current gain, typically 150.

DIFFUSED TRANSISTORS

Germanium diffused base transistor

The diffusion process offers better control of base width and junction area than the alloy process. For very high frequency transistors, one limitation can be the transit time of the minority carriers through the base region. In this connection the mobility of holes in n-type germanium is higher than in n-type silicon and so germanium offers higher frequency capabilities than silicon. Germanium diffused base transistors which operate up to several thousand megacycles per second are now available.

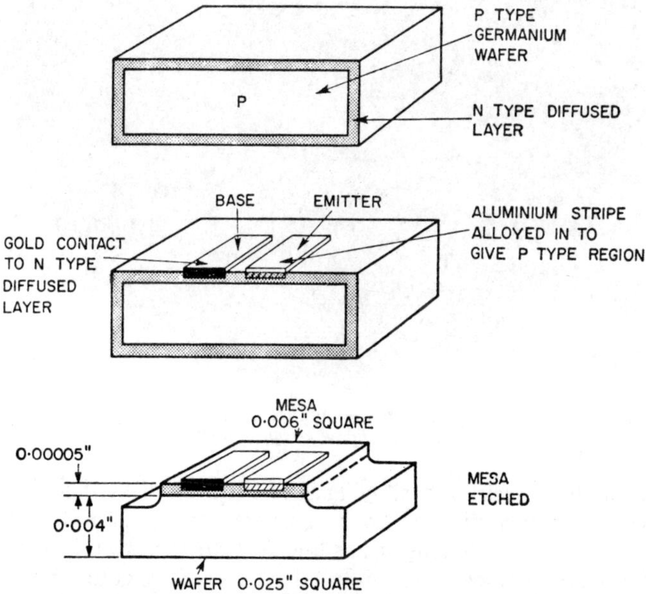

Fig. 4.8.—*Steps in the construction of a germanium diffused base transistor.*

The construction of a germanium diffused base transistor is as follows, the steps in the process being illustrated in Fig. 4.8. A thin n-type region is diffused into a p-type wafer by diffusing in a donor such as antimony. The depth of this diffused layer is of the order of 0·00005 inch. A very small area emitter junction is formed by evaporating aluminium on to the surface

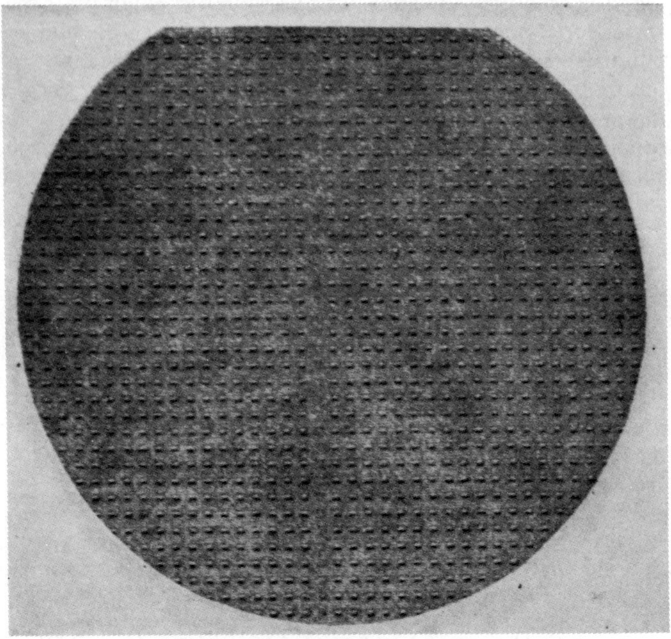

Fig. 4.9.—*A processed germanium slice containing over* 1,000 *diffused base transistor elements. Slice diameter approximately* 0.8 *inch.*

through a mask, and alloying it to form an alloyed junction with the n-type diffused layer. This emitter area is in the form of a stripe with dimensions as low as 0·001 inch by 0·003 inch. To make contact to the diffused base layer, a gold stripe of similar size is evaporated alongside the emitter, spaced about 0·0005 inch apart, and alloyed in at the same time. The final base width is about 0·00003 inch. The area of the diffused collector junction is defined by masking over the emitter and base stripes and etching the wafer to leave a "mesa" which contains the active part of the transistor. The operations so far described can all be carried out on a whole slice, including as many as 1,000 transistor elements, which, of course, are all

Fig. 4.10.—*Enlarged view of a single transistor element of Fig.* 4.9. *Base and emitter stripe dimensions are approximately* 0.006 *in* × 0.002 *in.*

processed simultaneously. A photograph of a typical processed slice is shown in Fig. 4.9. The slice is then cut up into separate transistor elements, as shown in Fig. 4.10, which are individually mounted on to headers. Contact is made to the aluminium emitter and gold base areas by thermocompression bonding a small diameter (0·0007 inch diameter) gold wire. Satisfactory bonds are made with a pressure of a few thousand pounds per square inch at a temperature about 300°C. A germanium diffused base transistor as described above will have a frequency cut-off of the order of 1,000 Mc/s.

Silicon diffused transistors

During the development of silicon diffused transistors several types have emerged in succession. Early types were formed by a simultaneous diffusion into n-type silicon of a

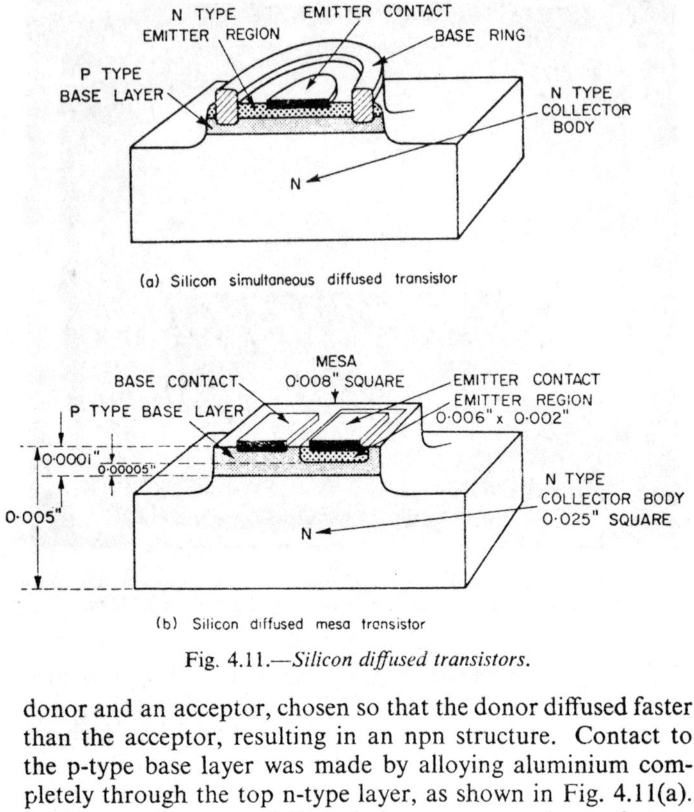

Fig. 4.11.—*Silicon diffused transistors.*

donor and an acceptor, chosen so that the donor diffused faster than the acceptor, resulting in an npn structure. Contact to the p-type base layer was made by alloying aluminium completely through the top n-type layer, as shown in Fig. 4.11(a).

This version was followed by a more advanced type (Fig. 4.11 (b)) in which the base and emitter layers were formed by separate acceptor and donor diffusions. After diffusing the p-type base layer, the surface of the silicon slice is oxidised, "windows" are etched in the oxide by a photo resist process (described more fully below), and the emitter formed by diffusing phosphorus through the window, the oxide surface preventing diffusion elsewhere. In both of these types the area of the collector junction is defined by "mesa" etching, and they are generally known as *diffused mesa transistors*. The collector–base junction at the surface of the mesa, although thoroughly cleaned by the etching and subsequent washing operations, is very susceptible to contamination which increases the collector leakage current and causes instability. Early investigations into methods of protecting the surface included surface oxidation over the junctions, and this led to the development of the so called *planar* process. This is a very important process and the steps in the process will be described fully. Many of the processes are similar to those used for diffused mesa transistors.

Silicon planar transistors

The steps in the planar process are illustrated in Fig. 4.12. As with other diffused transistors, the processes are carried out on the whole slice of silicon, a large number of transistor elements being processed simultaneously. For convenience, the illustrations in Fig. 4.12 show only one element.

An n-type silicon slice is first oxidised to a thickness about 0·00004 inch by heating to a temperature of 1,000°C in a flow of steam. The oxide is then removed from selected areas in preparation for the base diffusion by a process of photomasking. The surface of the oxidised slice is covered with a photo-sensitive emulsion, which when exposed to ultra-violet light is rendered insoluble and in that form resists attack by acid (photo-resist). The emulsion is illuminated through a photographic mask so that all of the wafer except the base areas is exposed to ultra-violet light; it is then developed and the unexposed areas washed away. Then the oxide on the silicon in these areas is dissolved out by immersing the slice in a hydrofluoric acid solution, the photo-resist protecting the oxide over the rest of the slice. The photo-resist material is finally removed from the surface and the slice cleaned ready for the base diffusion.

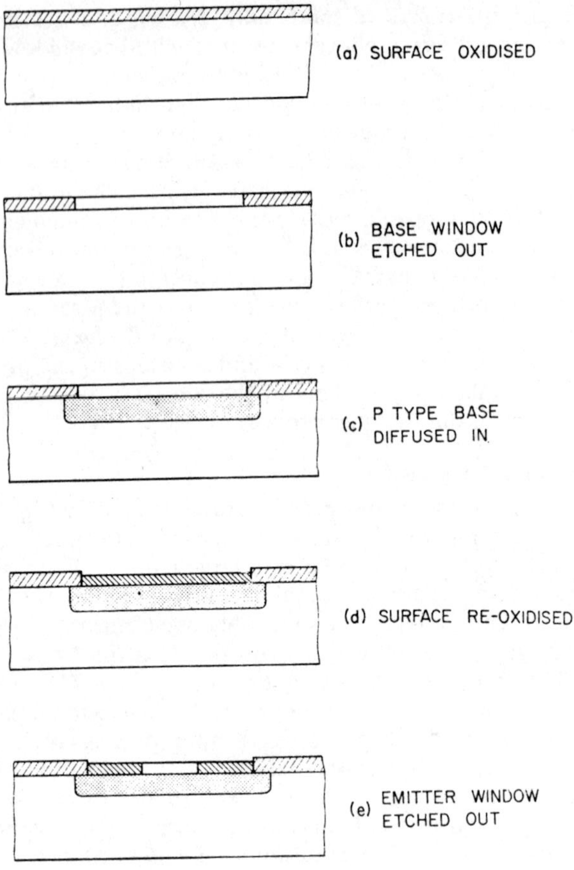

Fig. 4.12.—*The silicon diffused planar process. See over page for continuation.*

To form the base region, boron is diffused in through the "windows" in the oxide, the oxide preventing diffusion elsewhere. Boron tribromide is vapourised and the vapour mixed with a carrier gas such as nitrogen which is passed over the slices heated to a temperature of 850°C. Boron is deposited on

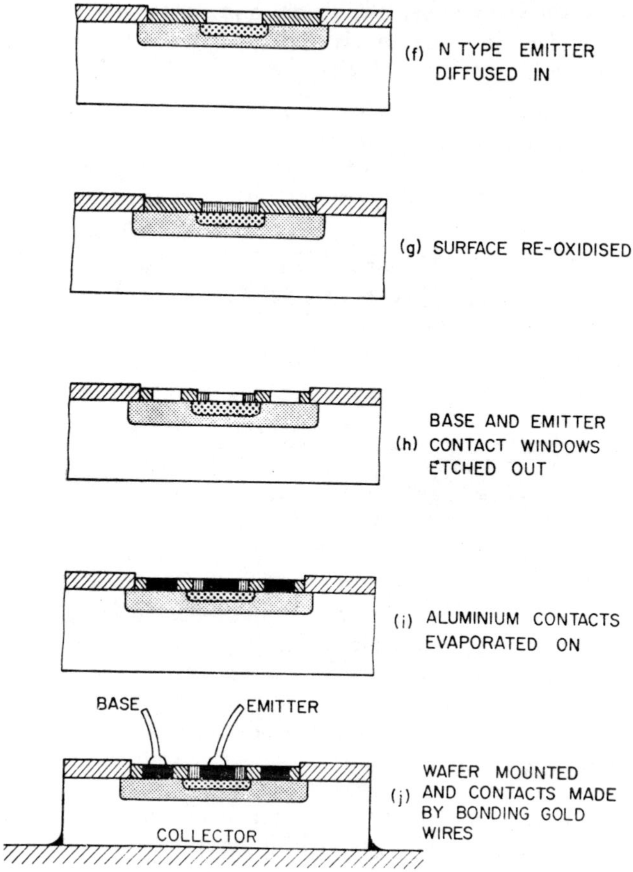

Fig. 4.12 (continued).—*Silicon diffused planar process.*

to the silicon surface and is then diffused into the silicon to the required depth by heating the slices at 1,200°C for about an hour, and finally a new oxide is formed over the diffused area by heating the slice in steam or oxygen. During the diffusion, in addition to moving directly into the silicon, the boron also diffuses sideways under the oxide, and so the pn junction forms under the oxide and is thus protected against surface contamination. A typical base diffusion depth is 0·00015 inch.

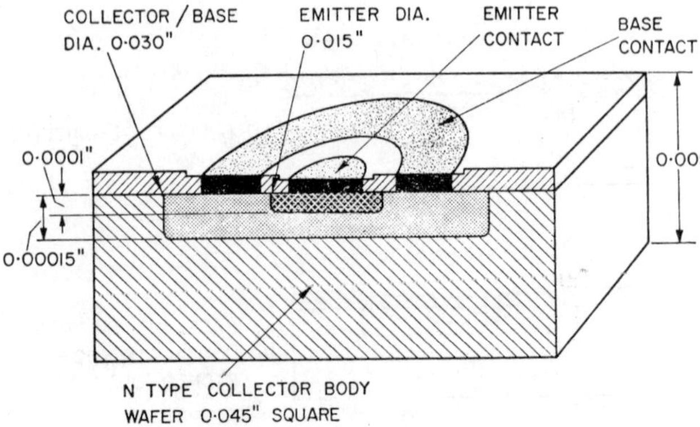

Fig. 4.13.—*Silicon planar transistor wafer.*

The slice is now prepared for emitter diffusion by preparing "windows" in the new oxide covering the base areas, using the same photo-resist process. To form the n-type emitter, phosphorus is diffused through the new windows in the oxide into the base region by heating phosphorus pentoxide in a flow of nitrogen which passes over the slice heated to 1,000°C. During the diffusion, the gas is changed to oxygen so that a final oxide is grown over the emitter region. The emitter diffusion depth is about 0·0001 inch, resulting in a base width between emitter and collector of 0·00005 inch.

The final process on the slice is to make metallised contact areas on the surfaces of the base and emitter regions. Again the photo-resist process is used to remove the oxide from suitable areas and aluminium is evaporated on to the exposed silicon and subsequently alloyed in. Fig. 4.13 shows a dimension drawing of a typical planar transistor element suitable for collector currents up to 200 milliamps with frequency cut-off in the 50–100 Mc/s region. Fig. 4.14 shows a photograph of a full slice after processing and Fig. 4.15 an enlarged view of a single transistor element.

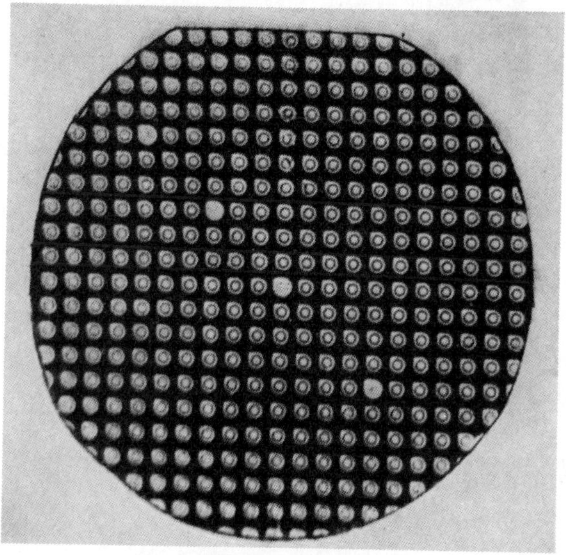

Fig. 4.14.—*A processed silicon slice containing over* 300 *planar transistor elements. Slice diameter approximately* 0.8 *inch.*

The slice is now cut into the individual elements by scribing between them and breaking. The individual wafers are assembled by fusing down to a header which forms the collector contact, and contact is made to the base and emitter regions by compression bonding a small diameter gold wire. A cut-away illustration of a typical unit is shown in Fig. 4.16.

The great advantage of the planar process is the protection of the collector–base junction at the surface by the oxide layer. This results in very low collector leakage current—for a unit with dimensions indicated in Fig. 4.13, the collector leakage current is of the order of 1 nanoampere at 10 volts and with a smaller unit having a collector diameter about 0·010 inch a

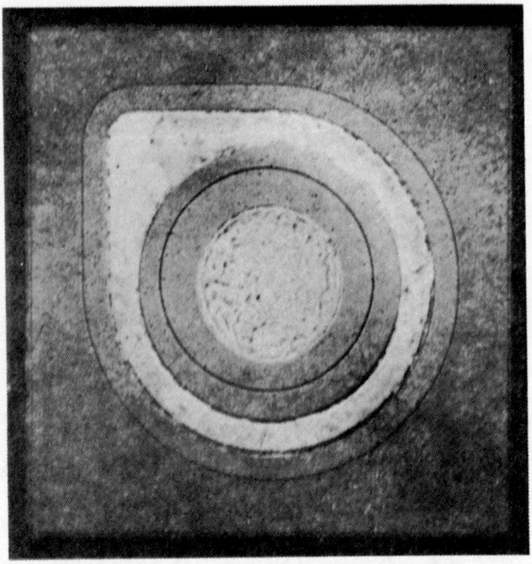

Fig. 4.15.—*Enlarged view of a single element in Fig.* 4.14. *Wafer dimension* 0.045 *inch square.*

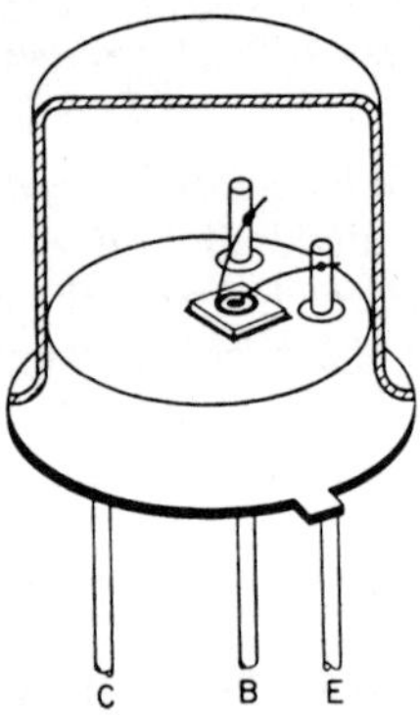

Fig. 4.16.—*Cut-away view of a silicon planar transistor.*

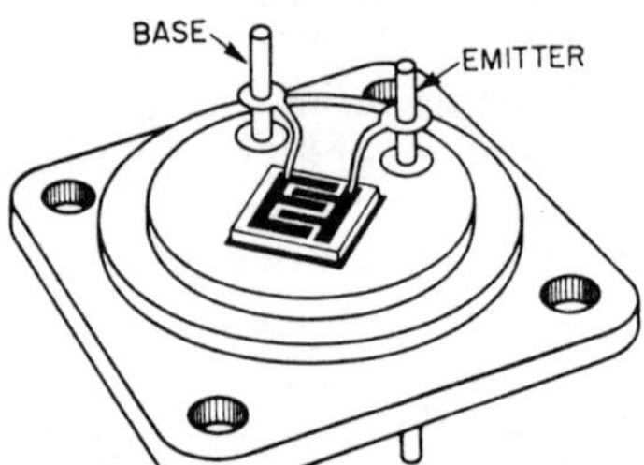

Fig. 4.17.—*Silicon diffused power transistor.*

typical leakage current is 0·1 nanoampere. Of equal importance is the fact that this low leakage is very stable due to the insulating oxide layer over the surface.

As with all diffused transistors, very thin base layers are possible, and this gives the possibility of very high frequency operation—up to several thousands of megacycles.

Another important advantage is that different size transistors can be made by substituting different photographic masks in the photo-resist process and so, with the same basic process and equipment, a whole range of types can be produced from a small signal, high frequency type (200 mW 600 Mc/s) to a high power medium frequency type (50 watts 10 Mc/s).

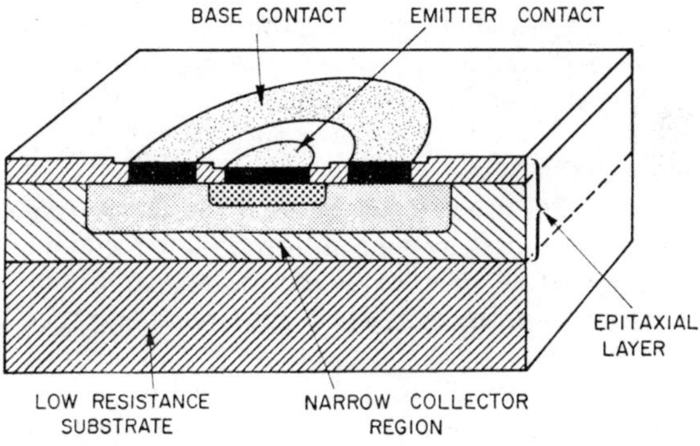

Fig. 4.18.—*Silicon epitaxial planar transistor.*

Although there are a multiplicity of steps in the planar process, these are carried out on the whole slice, many transistor elements being processed simultaneously, and so the final cost per transistor is not high.

By increasing the collector and emitter areas higher power transistors can be fabricated. One of the limitations in increasing the junction area is that the base current must flow out of the base region parallel to the junctions, and in doing so sets up a potential drop that causes the transistor action to be confined to an area adjacent to the periphery of the emitter. Thus to increase the effective area a multiplicity of long narrow emitter regions are used, with base contacts between them. This is illustrated in Fig. 4.17 which shows the structure of a typical silicon power transistor suitable for collector currents up to 5 amperes.

The main shortcoming of the diffused transistor is the series resistance between the collector junction and contact at the bottom surface of the wafer. The collector current must flow through this resistance and in doing so sets up an internal voltage drop, which results in a high value of saturation voltage. A considerable improvement can be obtained by using a composite slice consisting of a very low resistance substrate, on which has been deposited a thin epitaxial layer in which the transistor is formed. This is illustrated in Fig. 4.18. A planar transistor produced in this way is called an *epitaxial planar transistor*, and is a most versatile transistor with good all round characteristics.

Epitaxial transistors

The epitaxial process can be used for fabricating complete transistor structures by depositing alternate p and n layers on a suitable substrate slice. This can be accomplished by introducing first an acceptor into the gas stream and then changing over to a donor.

For one type of silicon epitaxial transistor a low resistivity p-type slice forms the collector. First an n-type layer is grown to form the base region, by introducing a donor into the hydrogen carrier gas with the silicon tetrachloride. The slice is then oxidised, and holes etched into the oxide as described earlier for planar transistors. Then a p-type epitaxial layer is grown on the silicon in the holes in the oxide by introducing an acceptor into the silicon tetrachloride. From this point, contact areas are laid down and the slice scribed up into individual transistor elements which are mounted as for planar transistors.

5
ASSOCIATED SEMICONDUCTOR DEVICES

In addition to the transistor, there are a number of other devices depending upon the properties of the pn junction. This family of semiconductor devices is quite extensive and is continually growing as new devices are developed. The transistor, in its various forms, is doubtless the most important, but many associated devices play important supporting roles in electronic circuit applications. Semiconductor power rectifiers and controlled rectifiers are used extensively for rectifying alternating voltage for d.c. power supplies, and diodes are used for many circuit functions such as detection, limiting and switching. The photosensitivity of a pn junction is utilised in photodiodes and phototransistors. Special types of transistor such as field effect transistors and unijunction transistors are proving valuable in adding to the scope of transistor application.

It would not be realistic to consider applications of the transistor without some knowledge of these associated devices, and so this chapter will be devoted to a review of such devices both old and new.

Semiconductor diodes and rectifiers

Basically, a diode is any rectifying device with two electrodes. It has become customary to use the word diode when the device is primarily intended for general circuit functions, and the word rectifier when the device is used to rectify power supplies—however small.

Point contact diodes

The point contact diode was the forerunner of the whole family of semiconductor devices. It was a formal development of the "crystal and catswhisker" detector used in the early days of radio broadcasting, and consists of a piece of semiconductor material—silicon or germanium—with a pointed metal wire "whisker" pressing on to its surface. Typical whisker materials are tungsten and platinum.

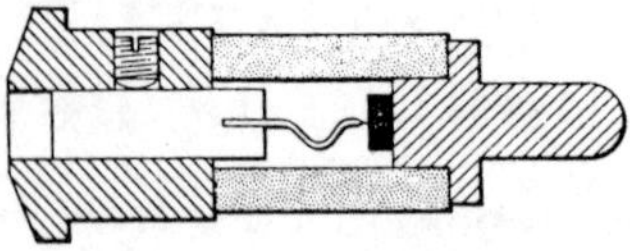

(a) SILICON CARTRIDGE MICROWAVE DIODE

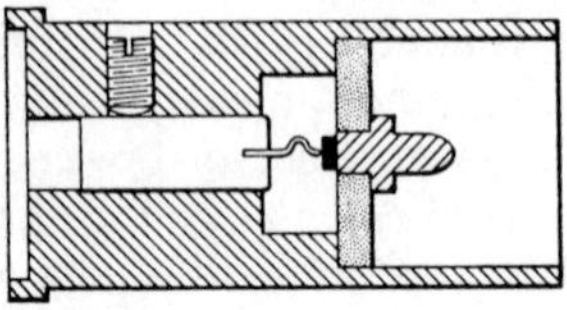

(b) SILICON CO-AXIAL MICROWAVE DIODE

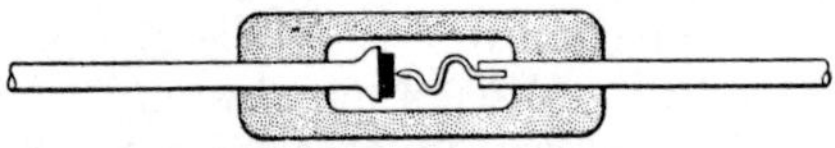

(c) GERMANIUM POINT CONTACT DIODE

Fig. 5.1.—*Point contact diodes.*

The silicon point contact diode, shown in Fig. 5.1 (a) and (b), is used as a frequency changer (mixer) or detector in the microwave frequency bands—mainly for radar and point-to-point communication systems with operating frequencies in the 3,000–30,000 Mc/s region. The germanium point contact diode illustrated in Fig. 5.1 (c) is useful for frequencies up to 100 Mc/s. It is available with voltage ratings up to 100 volts and current ratings of about 20 mA. It is, however, now being rapidly superseded by high frequency pn junction diodes.

Gold bonded diode

This is a variation of the germanium point contact diode in which a thin gold wire, 0·0015 inch diameter, is fused to the surface of an n-type germanium wafer by the passage of a current pulse. The gold includes a small percentage of gallium or indium (acceptors) which, during the fusing, alloys with the gold into the germanium to form a rectifying contact with characteristics similar to those of a pn junction diode. The gold bonded diode found wide application in computer circuits because of its combination of very low forward voltage drop (0·5 volt) with its high impedance reverse characteristic (leakage currents a few microamperes at −10 volts). As with the point contact diode, it is now being replaced by small area pn junction diodes.

Junction diodes

Small area pn junction diodes are usually made of silicon and are fabricated by diffusion techniques, using either the mesa or planar process. For general purpose types, a junction diameter of 0·006 inch is typical, giving a forward volt drop of

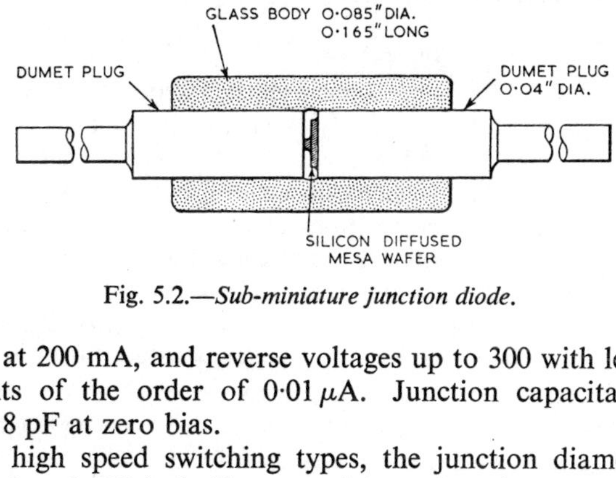

Fig. 5.2.—*Sub-miniature junction diode.*

1 volt at 200 mA, and reverse voltages up to 300 with leakage currents of the order of 0·01 μA. Junction capacitance is about 8 pF at zero bias.

For high speed switching types, the junction diameter is reduced to 0·003 inch diameter giving a junction capacitance of 2 pF and a switching time—the time the diode takes to change a current of 10 mA to zero—of 0·004 μsec.

These junction diodes are now being made with very small overall dimensions—a typical construction being shown in Fig. 5.2.

Varactor diode

The varactor diode is an extremely small area junction diode designed for use at microwave frequencies. Its application uses the fact that the junction capacitance varies inversely with reverse voltage, the main applications being in harmonic generators and parametric amplifiers. To make full use of the capacitance variation, it is necessary to reduce the internal series resistance of the diode to as low a value as possible.

The majority of varactor diodes are made from silicon using diffusion techniques. A typical diode has a junction diameter about 0·001 inch with a capacitance at zero bias of 1 pF and a series resistance of 2·5 ohms, giving a cut-off frequency (the frequency at which the junction capacitive reactance equals the series resistance) of 60 Gc/s. The junction capacitance varies approximately as the inverse square of the reverse voltage.

Varactor diodes are also made of gallium arsenide with cut-off frequencies up to 200 Gc/s.

Tunnel diode (Esaki diode)

The tunnel diode uses a pn junction with heavily doped p and n regions, giving a very abrupt junction with a very thin barrier. The important feature of the diode is that it has a negative resistance region, as shown in Fig. 5.3. In the forward direction, the first conduction mechanism is by "tunnelling" of majority carriers through the very thin junction barrier region. This tunnelling current dies away to be followed by the normal forward current due to minority carrier injection across the junction. Between these two components of current, there is a negative resistance region, the current decreasing as the voltage is increased. This characteristic can be used to give a very fast switching action. If a diode is initially biased at point A, a small current pulse will take it over the peak into the negative resistance region when it will rapidly move through the unstable region to the point B which is again a stable operating point. Thus the diode will "switch" from 0·07 volt to 0·38 volt.

With germanium the voltage between the "peak" and "valley" of the characteristic is about 0·3 volt and with gallium arsenide about 0·7 volt. Because of the heavy doping there is no high resistance region in the reverse direction. Since the initial current is by means of "majority" carrier tunnelling, the

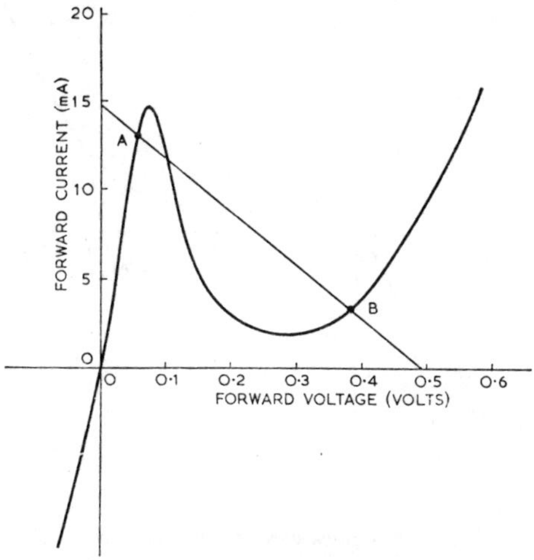

Fig. 5.3.—*Typical tunnel diode characteristic.*

switching speed can be very high, with switching times of less than one nanosecond (10^{-9} second), and so the diode can operate into the microwave region.

Backward diode

If the p and n regions of a silicon junction diode are heavily doped, but not sufficiently to produce the negative resistance region of the tunnel diode, the junction can have a characteristic as shown in Fig. 5.4, that is a normal forward characteristic with a "turn-up" at 0·7 volt but with a reverse leakage current increasing sharply from zero as with the tunnel diode. If viewed in a "backward" sense it can be regarded as a diode with a reverse breakdown of 0·7 volt and a forward volt drop of the order of 0·1 volt. Thus for low voltage signals, less than 0·7 volt peak, it will give a higher rectification efficiency than a normal diode, as illustrated in Fig. 5.5.

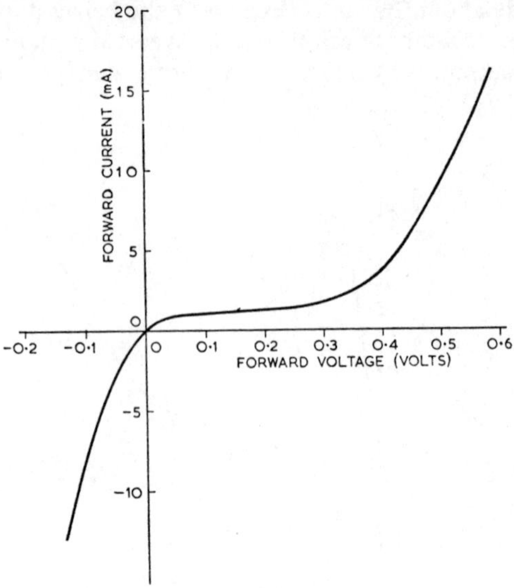

Fig. 5.4.—*Backward diode characteristic.*

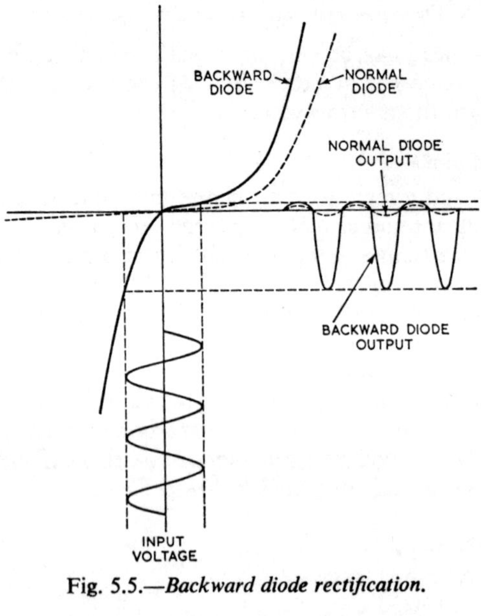

Fig. 5.5.—*Backward diode rectification.*

Junction rectifier

A semiconductor junction rectifier consists essentially of a pn junction mounted in such a way as to have low electrical resistance connections to the p and n regions of the junction, and also a low thermal resistance path from the junction to an

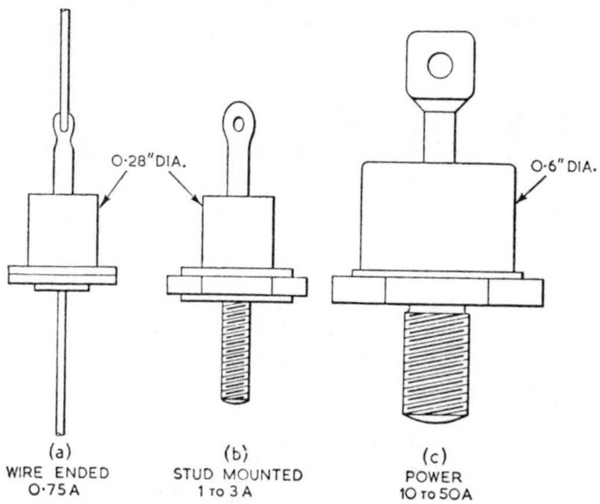

Fig. 5.6.—*Silicon rectifier outlines.*

external surface, so that a heat radiating system can be added to remove the heat generated by the power loss occurring in the unit. This internal power loss is equal to the product of the forward voltage and forward current, and because of the non-linear nature of the forward characteristic is approximately proportional to the forward current alone. There is also a small loss due to the reverse leakage current, but with silicon junction rectifiers this is normally negligible.

Compared with earlier forms of metal rectifier, semiconductor junction rectifiers are very efficient, due to their low forward voltage drop and high reverse voltage. Early junction rectifiers were made of germanium, but the vast majority are now made from silicon. There has been continual improvement over the last few years; silicon junction rectifiers are now usually made by the diffusion process and are available with reverse voltage ratings up to 1,500 volts and current ratings up to 200 amperes. Low current types, up to about one ampere, are made in the form of wire ended units and require no external heat radiating system. Above this current the units are made so that they can be bolted on to a fin system in order that the internally generated heat can be dissipated safely. Fig. 5.6 illustrates several typical silicon rectifier outlines.

Zener diode

The zener diode is a special version of the silicon junction rectifier with a well-defined sharp reverse breakdown at low voltage, usually in the range of 3–15 volts. The forward

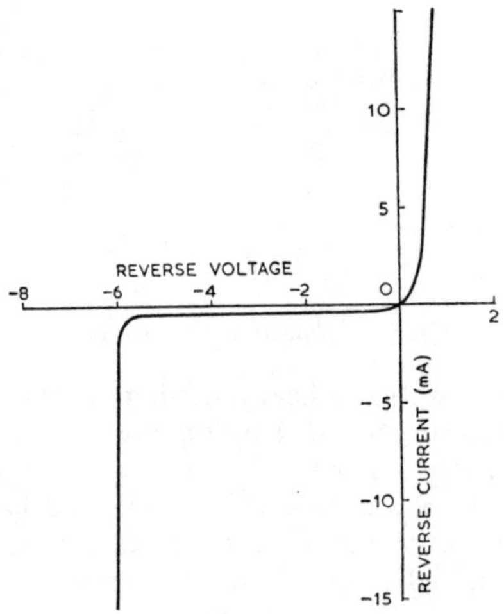

Fig. 5.7.—*Typical zener diode characteristic.*

characteristic is similar to that of the normal high voltage rectifier. The characteristics of a typical zener diode are shown in Fig. 5.7. In the reverse direction the leakage current is very low, but when the breakdown point is reached the reverse current increases very rapidly for a very small voltage increase, and so the device can be used as a voltage limiter. Alternatively, by biasing it into the breakdown region, it can be used to establish a stable voltage reference. The use of the zener diode in stabilised power supplies is described in a later chapter.

Silicon controlled rectifier (or thyristor)

The silicon controlled rectifier (scr), now in the U.K. officially named the thyristor, is a rectifying device which has two modes of operation. It normally blocks current flow in

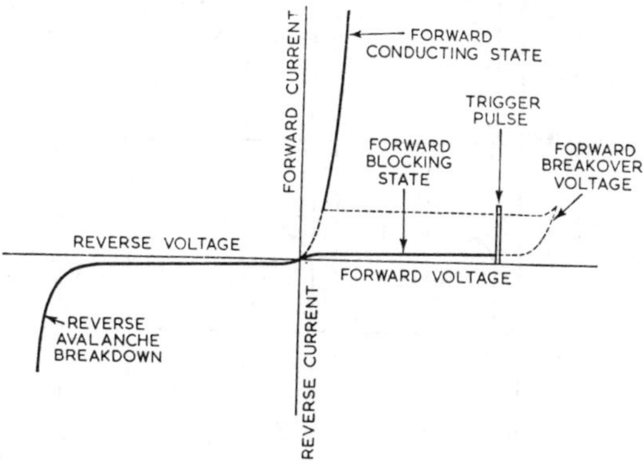

Fig. 5.8.—*Basic controlled rectifier characteristics.*

both directions, but it can be triggered so that it allows current flow in the forward direction while still blocking in the reverse direction. The electrical characteristics of the device are illustrated in Fig. 5.8. In the triggered state, it has characteristics similar to those of a conventional silicon rectifier. It is triggered into conduction by feeding a small current pulse into a third electrode, and once the device is conducting it will remain so until the load current is reduced almost to zero, when it automatically returns to the blocking state. The electrode to which the trigger pulse is fed is usually called the *gate* electrode.

The device can be used to control the output current of a rectifier circuit. Consider a sinusoidal voltage applied to the scr as shown in Fig. 5.9. As the voltage increases in the forward direction, no current flows until the triggering pulse is applied. Then load current flows and follows the sinusoidal voltage for the remainder of the half cycle. During the reverse half cycle

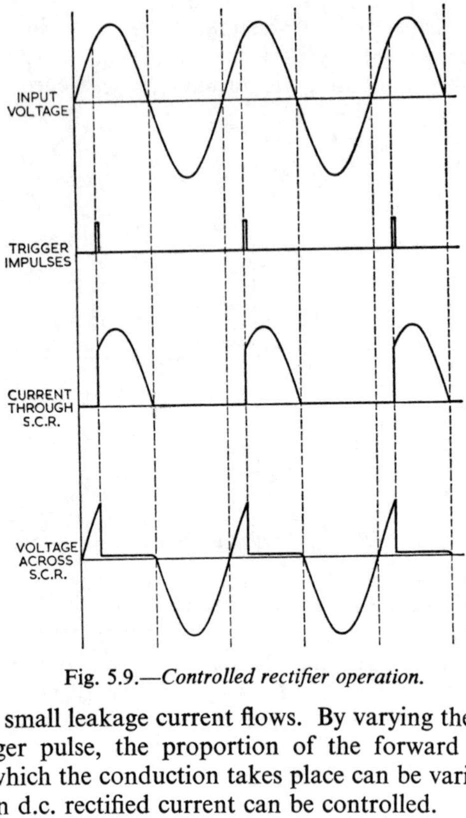

Fig. 5.9.—*Controlled rectifier operation.*

only the small leakage current flows. By varying the timing of the trigger pulse, the proportion of the forward half cycle during which the conduction takes place can be varied, and so the mean d.c. rectified current can be controlled.

Operation

The controlled rectifier consists basically of a piece of silicon with a PNPN structure as shown in Fig. 5.10. A simplified picture of its operation can be obtained by considering it as a pnp transistor (PNp) and an npn transistor (npN) coupled together with a common collector junction J_2. With the polarity shown, junction J_1 is biased forward (+ve on p side), J_2 reverse and J_3 forward. Thus J_1 is biased as an emitter

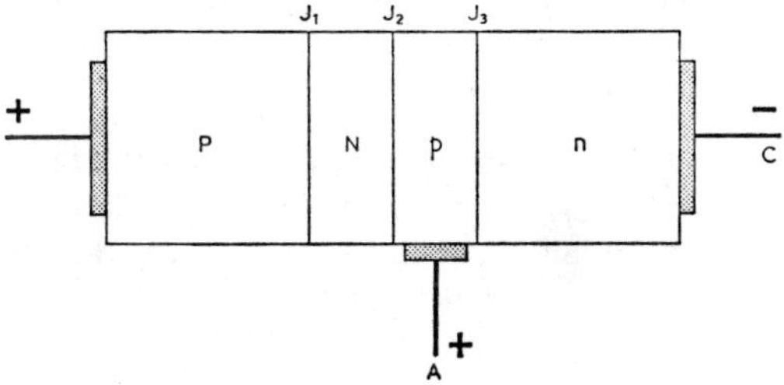

Fig. 5.10.—*Four-layer structure.*

junction for transistor PNp, J_3 as an emitter for transistor npN and J_2 as a reverse biased collector junction for both. Now consider the components of current passing through junction J_2. First, since it is reverse biased, there will be the small leakage current I_{CO}. Considering the PNp transistor, the current through the emitter junction J_1 is equal to I, the external circuit current. The current through the collector junction J_2 due to this emitter current I will equal $\alpha_1 I$, where α_1 is the forward current transfer ratio of the PNp transistor. Similarly we have an emitter current through J_3 equal to I and a collector current through J_2 of $\alpha_2 I$ where α_2 is the current transfer ratio of the npN transistor. Thus the total current through J_2 is $I_{CO}+\alpha_1 I+\alpha_2 I$. But the total current at any point in the circuit must be I. Thus

$$I = I_{CO}+\alpha_1 I+\alpha_2 I$$

or

$$I = \frac{I_{CO}}{1-(\alpha_1+\alpha_2)}$$

If the sum of α_1 and α_2 is low, I will be low—of the order of I_{CO}—and the device will have a high impedance, the reverse biased junction J_2 blocking the applied voltage. If the sum of α_1 and α_2 is increased towards unity, the current through J_2

increases, and eventually, when $\alpha_1 + \alpha_2$ equals unity, the current suddenly increases to a high value, which is only limited by the resistance of the load in the external circuit.

Thus if the sum of $\alpha_1 + \alpha_2$ is initially low and is then suddenly increased to unity, the device will trigger from the high impedance to the conducting state.

Triggering

To increase $(\alpha_1 + \alpha_2)$, advantage is taken of the fact that the forward current transfer ratio of a silicon transistor is small at low emitter current, and increases rapidly as the emitter current

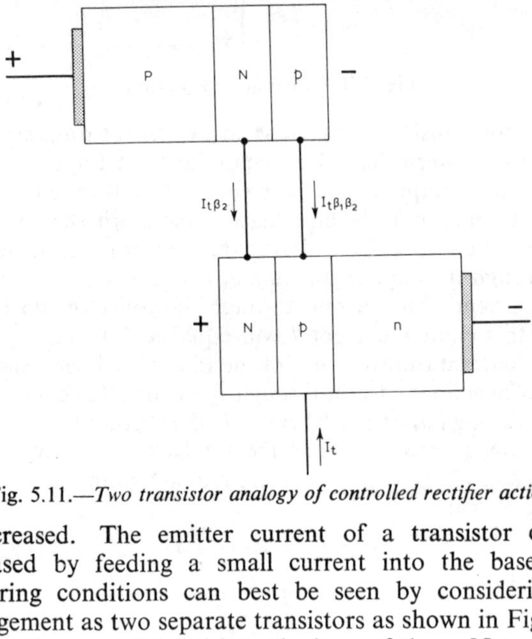

Fig. 5.11.—*Two transistor analogy of controlled rectifier action.*

is increased. The emitter current of a transistor can be increased by feeding a small current into the base. The triggering conditions can best be seen by considering the arrangement as two separate transistors as shown in Fig. 5.11. If a small current I_t is fed into the base of the npN transistor, it will be amplified, and appear at the collector as $I_t\beta_2$ where β_2 is the common emitter current gain of transistor npN. This current flows into the base of transistor PNp and is amplified again to $I_t\beta_2\beta_1$ where β_1 is the current gain of the PNp transistor. If $I_t\beta_2\beta_1$ is greater than I_t the system will be regenerative and the device will trigger. Thus the condition for trigger is $\beta_1\beta_2 > 1$. It should be remembered that the actual values of β_1 and β_2 will depend on the value of I_t.

Referring back to Fig. 5.10, the device can be triggered by causing a small current pulse to flow between A and C.

If no trigger pulse is applied and the applied voltage is steadily increased, the leakage current I_{CO} will increase and a point will eventually be reached where this increase of leakage current will result in sufficient increase of $\alpha_1 + \alpha_2$ to cause triggering. This is called *self-breakover* and sets a limit to the voltage that can be applied.

With the applied voltage in the opposite polarity, there is no possibility of regenerative build up. Junctions J_1 and J_3 are both biased in the reverse direction, and the overall characteristic is similar to that of a single pn junction, the voltage being limited by avalanche breakdown of the junctions.

Construction and characteristics

Silicon controlled rectifiers are usually manufactured by the diffusion process, starting with an n-type wafer into which an acceptor is diffused to give a pnp structure. Then a donor is

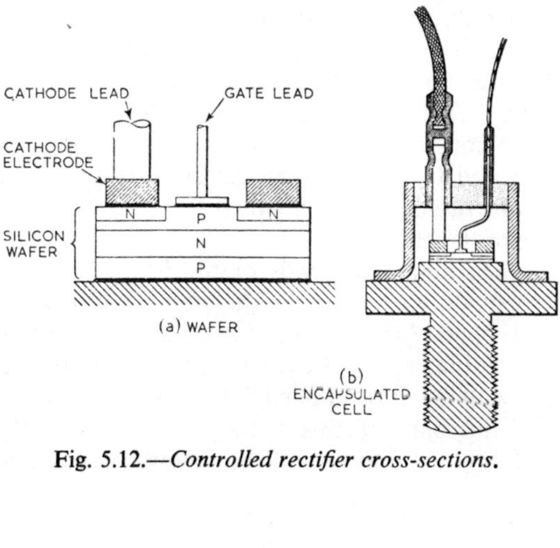

Fig. 5.12.—*Controlled rectifier cross-sections.*

diffused into one surface to give the required pnpn configuration. The thicknesses of the various regions are controlled to give the required voltage and trigger characteristics. Electrode contacts are made to the two end regions, and trigger contact to the centre n or p region.

Typical units will withstand up to 600 volts, and special types up to 1,500 volts, with current ratings from one or two amperes up to 100 amperers. Triggering currents are from 10 mA for the low current types to 50 mA for the high current types. Typical constructions are illustrated in Fig. 5.12.

Four layer diodes

The four layer diode is a small area two terminal version of the controlled rectifier, designed for use in switching circuits. No trigger contact electrode is provided, the device relying on self-breakover for switching into conduction. The diode is usually designed to have self-breakover within a specified small range of voltage. A typical unit will switch into conduction with an applied voltage of 65 volts $\pm$ 5 volts and has a current rating about 100 mA.

The unijunction transistor

The unijunction transistor is a three terminal semiconductor device with only one pn junction. The arrangement is such that the device exhibits a negative resistance characteristic and so can be used for switching applications and oscillator circuits. It was originally called the *double base diode.*

The basic construction is shown in Fig. 5.13. Two ohmic contacts, B and C, are made to the ends of an n-type bar of silicon. A rectifying junction called the emitter, E, is formed on the side of the bar between B and C. In operation, a voltage V_{BC} is applied across the ends of the bar. With no current flowing through the rectifying junction, the bar acts as a straight resistance (about 5,000 ohms) and a fraction of V_{BC} will appear at the region of the bar opposite the emitter. Now consider a voltage V_{EC} applied between the emitter and the bottom of the bar. If this voltage is less than the above fraction of V_{BC} the emitter will be reverse biased, and only a

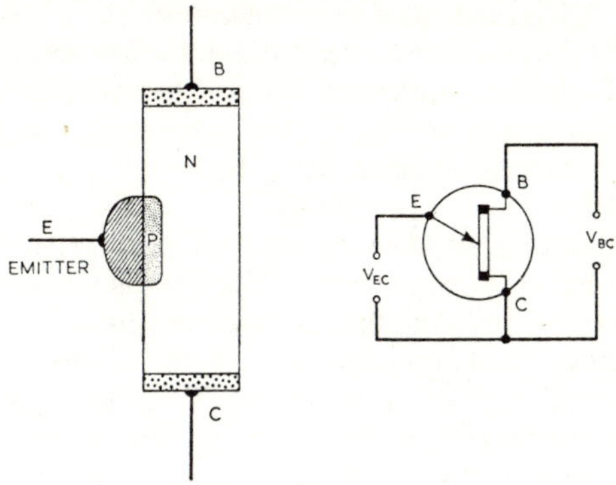

Fig. 5.13.—*Basic unijunction transistor.*

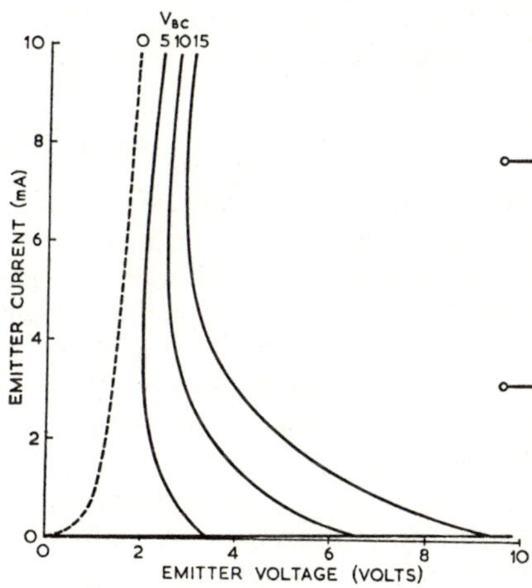

Fig. 5.14.—*Unijunction transistor characteristics.*

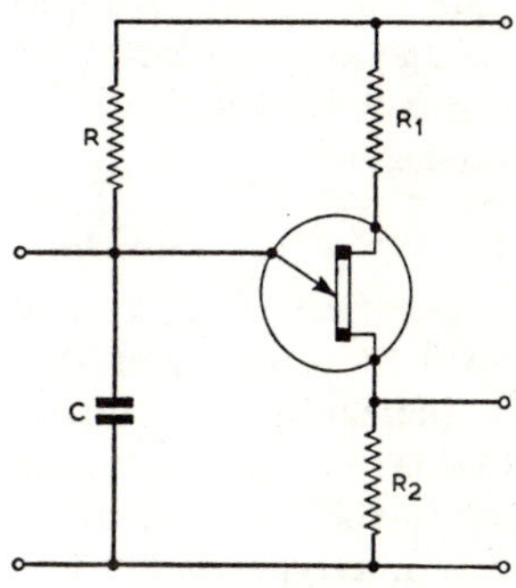

Fig. 5.15.—*Use of a unijunction transistor as a relaxation oscillator.*

small reverse leakage current will flow. If V_{EC} is increased to become greater than the fraction, the emitter will become forward biased, and current will flow in the form of holes emitted into the bar. These holes move down to C and their presence reduces the resistance of the bar between E and C resulting in a decrease of the fraction of V_{BC} appearing opposite the junction in the bar. Thus as the emitter current increases, the voltage between E and C decreases, giving the negative resistance characteristic shown in Fig. 5.14. Dotted in the same diagram is the characteristic of the emitter diode with no voltage (V_{BC}) applied across the ends of the bar.

The negative resistance can be utilised for a number of special circuits. A good example is the relaxation oscillator shown in Fig. 5.15. The operation of the circuit is as follows. The capacitor C charges up through the resistor R until V_E reaches the value at which the emitter starts to conduct. Then, the capacitor rapidly discharges through the device and the low value load resistance R_2. When the capacitor has completely discharged, the unijunction transistor returns to its original state and capacitor C starts to charge up again. A sawtooth voltage is generated across C due to the charge and discharge, and a voltage pulse is generated across R_2 by the discharge current.

The field effect transistor

The field effect transistor consists essentially of a bar of semiconductor material, the resistance of which can be varied by increasing the reverse bias across a pn junction on the side of the bar and so expanding the junction depletion region across the bar. The arrangement is illustrated in Fig. 5.16. A bar of p-type silicon has n-type regions formed on the two opposite faces of the bar. The n-type regions are called *gates* and the space between the gates is called the *channel.* The two ends of the bar are called the *source* and *drain* respectively. The two n-type regions are connected together externally, and if a voltage is applied between them and the source electrode, so as to bias the pn junctions in the reverse direction, a depletion region will be formed in the p-type channel on each side. As the voltage is increased, the two depletion layers will widen and extend towards each other, thus reducing the effective width of the channel and increasing the resistance between the two ends of the bar. The channel is wedge shaped due to the voltage gradient down the bar.

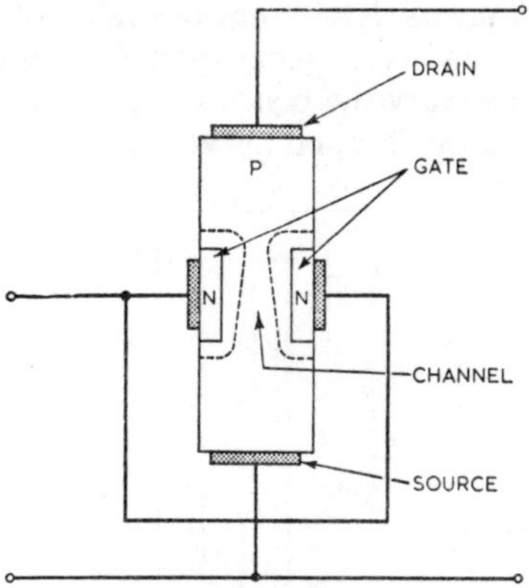

Fig. 5.16.—*Basic field effect transistor.*

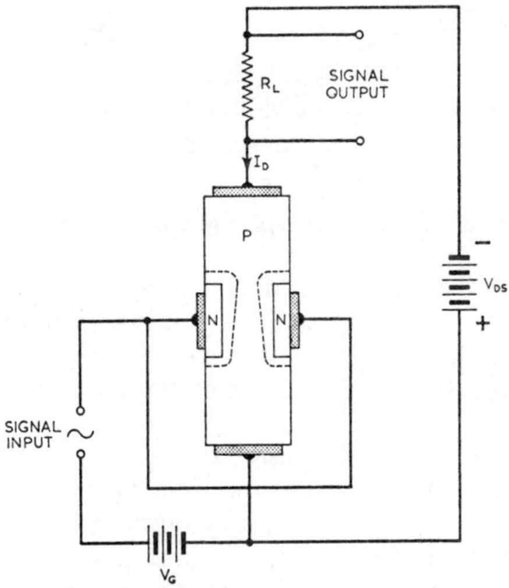

Fig. 5.17.—*Field effect transistor operation.*

In operation, a voltage V_{DS} is applied across the ends of the bar through a load resistance R_L as shown in Fig. 5.17. A bias voltage V_G is applied to the gate, and the resulting penetration of the depletion layers determines the value of the current I_D flowing down the bar. If a signal voltage is applied to the gate as shown, this will vary the penetration of the depletion layer and so produce a corresponding variation in the current I_D.

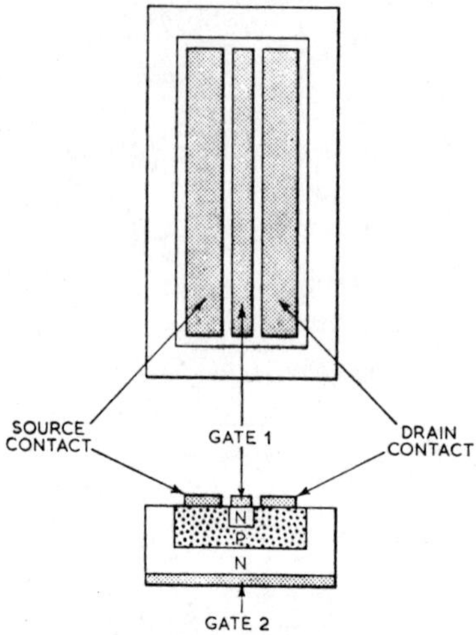

Fig. 5.18.—*Diffused field effect transistor construction.*

This change of current will produce an amplified voltage across the load resistance R_L. Since the signal is applied to a reverse biased junction, the input impedance is very high—with silicon planar field effect transistors it can be 10 MΩ or higher. Thus this type is suitable for many applications where the input resistance of the normal transistor is too low.

Another most important feature of the field effect transistor is that only one type of carrier is involved in the operation—the majority carrier in the bar of semiconductor. This is in distinct contrast to the junction transistor, which depends upon the flow of minority carriers through the base region and majority carriers in the emitter and collector regions.

The construction of a diffused field effect transistor is shown in Fig. 5.18. The unit is made by the planar process described earlier. An n-type wafer has a p-type region diffused in to form the channel and then a very narrow n-type region is diffused centrally as shown to form one gate. Contacts are made to the p-type region on each side of the narrow gate to form

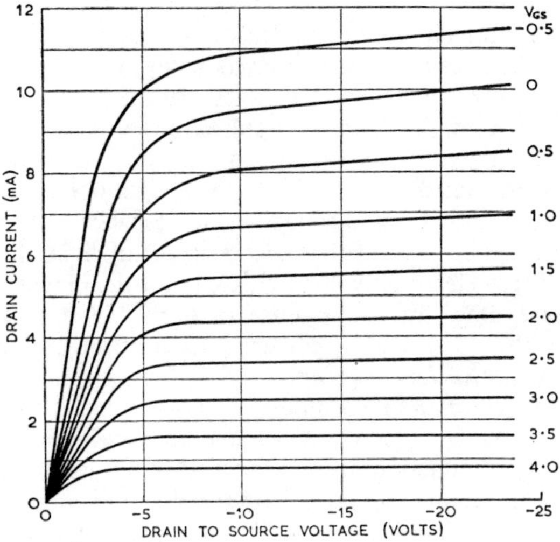

Fig. 5.19.—*Field effect transistor characteristics.*

the drain and source electrodes. The n-type wafer forms the second gate extending for the whole distance between source and drain.

A typical family of characteristics of a field effect transistor are shown in Fig. 5.19. These curves show the variation of drain current with variation of drain to source voltage for several values of gate to source bias voltage. Typical parameters are:

Static source to drain resistance	1,000 ohms
Drain to gate breakdown voltage	25 volts
Mean operating drain current	5 mA
Transfer conductance	2 mA/volt

In addition to the high input resistance, the field effect transistor has a low noise factor and is very useful for high impedance, low noise, input stages.

Photodiodes

When light falls on a pn junction, the light energy is absorbed into the crystal lattice and increases the energy level of the electrons. This causes some to break away from their atoms, thus creating free electrons and holes. In the *photoconductive* photodiode, a reverse voltage is applied to the pn junction, and the holes and electrons created near the junction by the light energy are attracted across it in opposite directions to form a current which flows out through a load resistance to produce an output signal. The higher the intensity of light the more holes and free electrons will be created and the higher will be the current flow. In the absence of light, the only current flowing is the small reverse leakage current of the pn junction. This is called the *dark current*.

In the *photovoltaic* photodiode no bias voltage is applied to the diode. As was described in Chapter 2, with no voltage across the pn junction free carriers will diffuse across the junction and set up a potential barrier. When light falls on the junction the density of carriers diffusing across the junction will increase. If the region on one side of the junction is made very thin it can become saturated with carriers and then a differential voltage is set up across the junction. This voltage can cause a current to flow in a resistance connected across the terminals of the device. The most common example of the photovoltaic diode is the *solar cell* which converts energy received from the sun into electrical energy. In bright sunlight the output from typical solar cells is of the order of 60 watts per square meter of cell surface area.

Phototransistor

The phototransistor is a junction transistor constructed so that light can fall on the structure near the collector junction. The action of the light increases the current through the reverse biased collector junction and, if the base terminal of the transistor is left open circuit, the increase of current is amplified by the transistor to result in a higher collector current, in much the same way as the collector/base leakage current is amplified in the common emitter connection. One of the important features of photoconductive diodes and phototransistors is the very small active area of the devices. This enables them to be used very successfully for position indicators. Also, if incident light is focused from a large area on to the active region, very high sensitivities can be obtained.

6
TRANSISTOR EQUIVALENT CIRCUITS AND PARAMETERS

IN circuit design associated with the application of a transistor it is helpful to represent the transistor by an electrical equivalent consisting of ordinary passive components, resistors and capacitors, together with an active generator which simulates the amplifying properties of the transistor. Such equivalent circuits are useful in two ways—they enable the manufacturer to specify his device in a manner that will be useful to the circuit design engineers, and also they enable the design engineer to carry out circuit evaluation to help decide the best arrangement for any particular application.

Simple T equivalent circuit

In developing an equivalent circuit, the first obvious approach is to look at the structure of the device. Considering the block representation of a transistor shown in Fig. 6.1

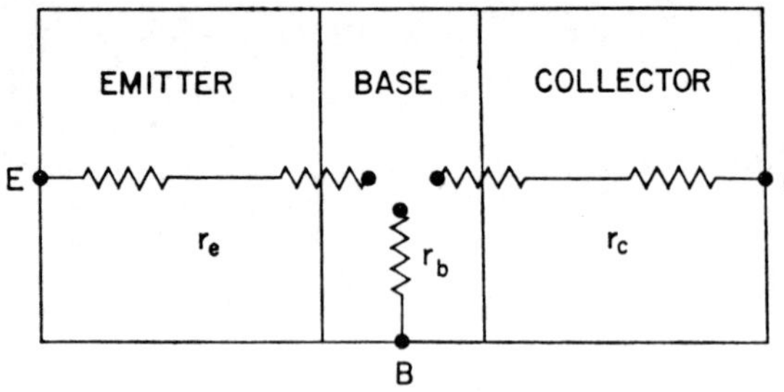

Fig. 6.1.—*Schematic representation of transistor internal resistances.*

certain features can be approximated. First, the emitter current, in flowing from the emitter terminal through the emitter junction, encounters a resistance r_e. This resistance consists of two components, the ohmic resistance of the emitter region, and the forward resistance of the emitter junction. In a properly designed transistor, the former is very low and can be neglected. Similarly the collector current, in flowing through the collector junction to the collector terminal, encounters a

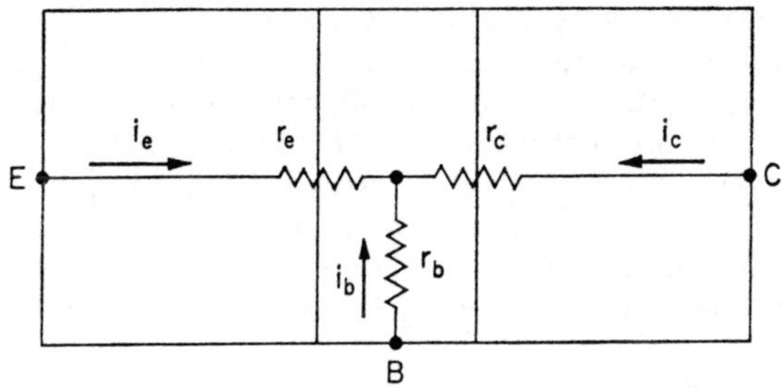

Fig. 6.2.—*Basic T arrangement of transistor internal resistances.*

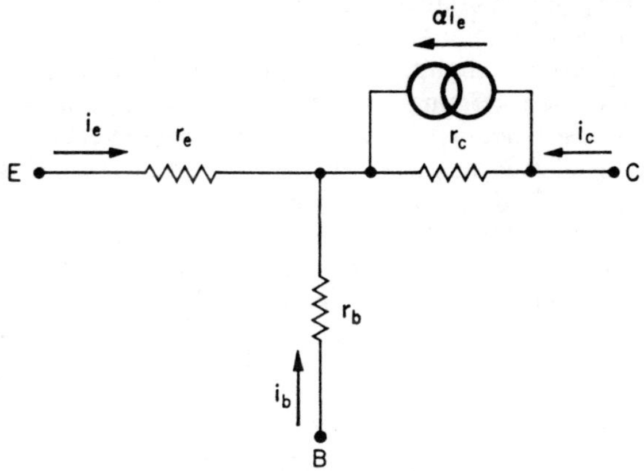

Fig. 6.3.—*Basic T equivalent circuit.*

resistance r_c consisting of the high resistance of the reverse biased junction and the ohmic series resistance of the collector region. Again, the latter can be neglected. The base current, in flowing into and along the narrow base region parallel to the junctions, meets a certain resistance r_b associated with the ohmic resistance of the base region. These three resistances are, of course, distributed over the cross section, but for a simplified picture it can be assumed that they are lumped components meeting at some point in the base region, as indicated in Fig. 6.2. This arrangement forms the basis of what is known as the *T equivalent circuit*.

All that remains to complete the circuit is to include a current generator in the collector circuit equal to αi_e, that is, the emitter current multiplied by the forward current transfer ratio, which represents the current flowing through the collector due to the emitter current i_e. This gives the final circuit shown in Fig. 6.3. To establish a standard notation, positive current flow is taken as flowing inwards from any terminal.

Resistance parameters

Certain properties of a transistor can be determined by measuring the resistance between emitter and base with the collector open circuit and between the collector and base with the emitter open circuit. If the equivalent circuit of Fig. 6.3 is assumed, these terminal measurements can be related to the internal resistances r_e, r_b and r_c. Such measurements would be small signal a.c. measurements, with the transistor biased to give the required d.c. operating conditions.

Referring to Fig. 6.4, measurements can be made as follows:

Input resistance with collector open circuit

(designated r_{ib}—resistance, *i*nput, common *b*ase)

$$r_{ib} = \frac{v_e}{i_e} = r_e + r_b$$

Output resistance with emitter open circuit

(designated r_{ob}—resistance, *o*utput, common *b*ase)

$$r_{ob} = \frac{v_c}{i_c} = r_c + r_b$$

The reverse internal feedback from output to input via the resistance network can be measured by applying a voltage at the collector and measuring the voltage produced across the open circuit emitter:

Feedback resistance with emitter open circuit
(designated r_{rb}—*r*esistance, *r*everse transfer, common *b*ase)

$$r_{rb} = \frac{v_e}{i_c} = r_b$$

These three resistances, r_{ib}, r_{ob} and r_{rb} are called the open circuit resistance parameters and together with the forward current transfer ratio α (the ratio of collector current to emitter current with the collector short circuited) they define the small signal properties of a transistor for common base operation in low frequency applications.

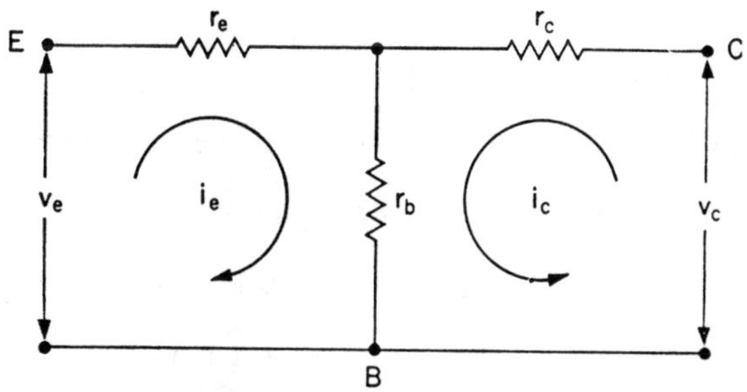

Fig. 6.4.—*Measurement of resistance parameters.*

Common emitter T equivalent circuit

The T equivalent circuit can be applied to the common emitter arrangement as shown in Fig. 6.5 by transposing the emitter and base resistors. In using this equivalent circuit, it should be noted that the current generator is still in terms of the emitter current i_e and the common base current transfer ratio α. The values of the current generator and collector resistance can be modified so that they are expressed in terms of base current and common emitter current gain as indicated in Fig. 6.6, but there are some points in favour of retaining the values in Fig. 6.5 in view of their basic nature.

By way of illustration, values of r_e, r_b and r_c for a typical small signal germanium alloy transistor are 30 ohms, 400 ohms and 1 MΩ respectively. These values will be used in subsequent sections to obtain typical values for certain other features.

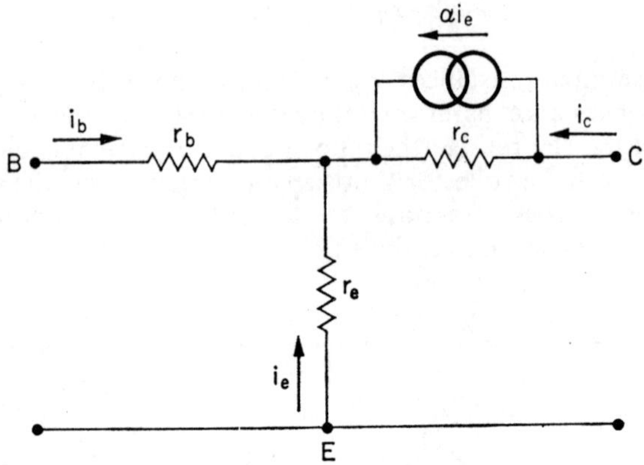

Fig. 6.5.—*Common emitter T equivalent circuit.*

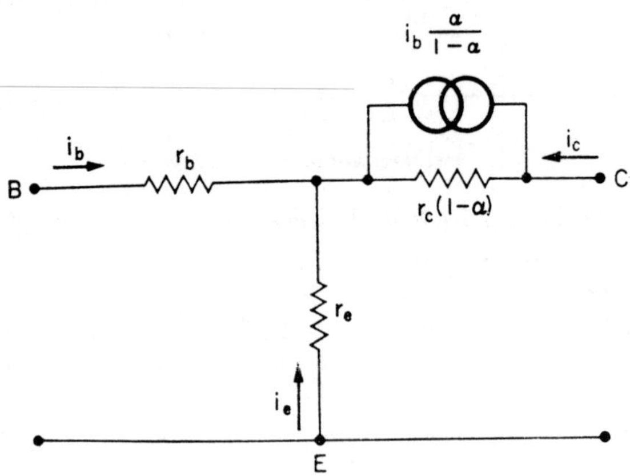

Fig. 6.6.—*Common emitter T equivalent circuit, alternative values.*

Hybrid parameters and equivalent circuit

Although the simple T equivalent circuit described above is quite a reasonable approach and is very useful in illustrating the evolution of an equivalent circuit, some difficulties are encountered in the measurements involved, and it has now become customary throughout the semiconductor industry to use a different set of parameters called the hybrid parameters (*h* parameters). Taking, for example, the common emitter arrangement, the *h* parameters are defined as:

h_{ie}—*h* (parameter), *i*nput resistance, common *e*mitter, output shorted.

h_{re}—*h* (parameter), *r*everse voltage feedback ratio, common *e*mitter, input open.

h_{fe}—*h* (parameter), *f*orward current gain, common *e*mitter, output shorted.

h_{oe}—*h* (parameter), *o*utput admittance, common *e*mitter, input open.

It will be seen that this set of parameters consists of one resistance, one admittance, one voltage ratio and one current ratio—hence the term hybrid.

The common emitter equivalent circuit using these parameters is shown in Fig. 6.7. In this circuit the forward current transfer ratio (current gain) h_{fe} appears as a current generator

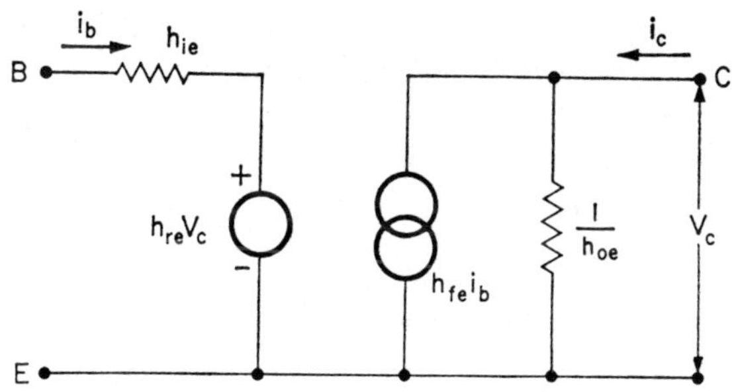

Fig. 6.7.—*Common emitter hybrid equivalent circuit.*

in the output circuit as in the T equivalent circuit. The reverse voltage feedback ratio h_{re} appears as a voltage generator in the input circuit.

If the values of one group of parameters are known for one equivalent circuit, and for one circuit arrangement (common

Table 6.1. Parameter conversion formulae

Symbol	*Typical value*	*Common base*	*Common emitter*	*Common collector*	T *Equivalent circuit*
α	0·98	$-h_{fb}$	$\frac{h_{fe}}{1+h_{fe}}$	$\frac{1+h_{fc}}{h_{fc}}$	α
r_e	30 Ω	$h_{ib}-\frac{h_{rb}(1+h_{fb})}{h_{ob}}$	$\frac{h_{re}}{h_{oe}}$	$\frac{1-h_{rc}}{h_{oc}}$	r_e
r_b	400 Ω	$\frac{h_{rb}}{h_{ob}}$	$h_{ie}-\frac{h_{re}(1+h_{fe})}{h_{oe}}$	$h_{ic}-\frac{h_{fc}(1-h_{re})}{h_{oc}}$	r_b
r_c	1 MΩ	$\frac{1-h_{rb}}{h_{ob}}$	$\frac{1+h_{fe}}{h_{oe}}$	$-\frac{h_{fc}}{h_{oc}}$	r_c
h_{ie}	1,900 Ω	$\frac{h_{ib}}{1+h_{fb}}$	h_{ie}	h_{ic}	$r_b+\frac{r_e}{1-\alpha}$
h_{fe}	49	$-\frac{h_{fb}}{1+h_{fb}}$	h_{fe}	$-(1+h_{fc})$	$\frac{\alpha}{1-\alpha}$
h_{re}	$1{\cdot}5\times10^{-3}$	$\frac{h_{ib}\,h_{ob}}{1+h_{fb}}-h_{rb}$	h_{re}	$1-h_{rc}$	$\frac{r_e}{r_c(1-\alpha)}$
h_{oe}	$0{\cdot}5\times10^{-4}$ mho	$\frac{h_{ob}}{1+h_{fb}}$	h_{oe}	h_{oc}	$\frac{1}{r_c(1-\alpha)}$

base, common emitter or common collector), the values of parameters for other equivalent circuits and circuit arrangements can be calculated. Table 6.1 gives the corresponding values for the hybrid and T equivalent circuits and the three circuit arrangements. All of the expressions relate to small signal a.c. values, measured at predetermined d.c. bias conditions.

Variation of parameters with operating conditions

All parameters are dependent to some extent on the actual operating values of collector current, collector voltage and junction temperature. In applying the equivalent circuits to

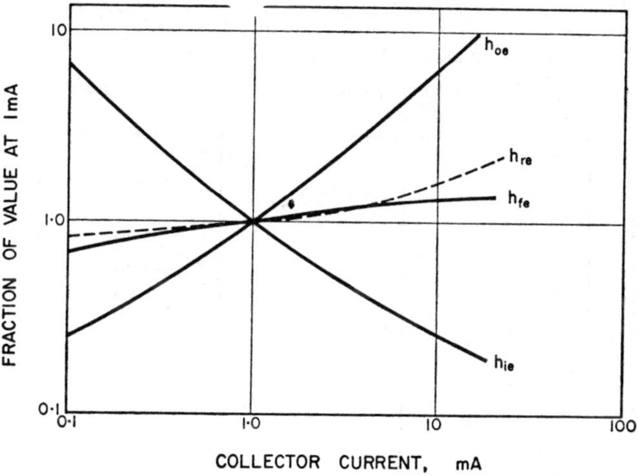

Fig. 6.8.—*Variation of h parameters with collector current.*

circuit design, it is important to correct the values of the parameters to conform with the anticipated operating conditions. The variation of parameters of most types of transistor will differ in detail, depending on their construction, but manufacturers provide full information on their devices.

For a typical small signal germanium alloy transistor the variation of the common emitter *h* parameters with collector

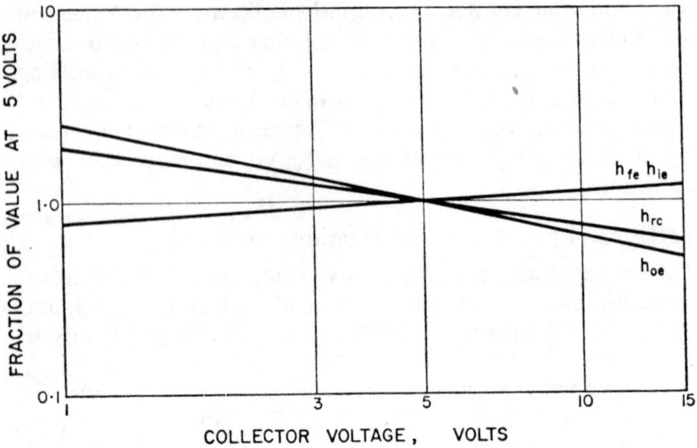

Fig. 6.9.—*Variation of h parameters with collector voltage.*

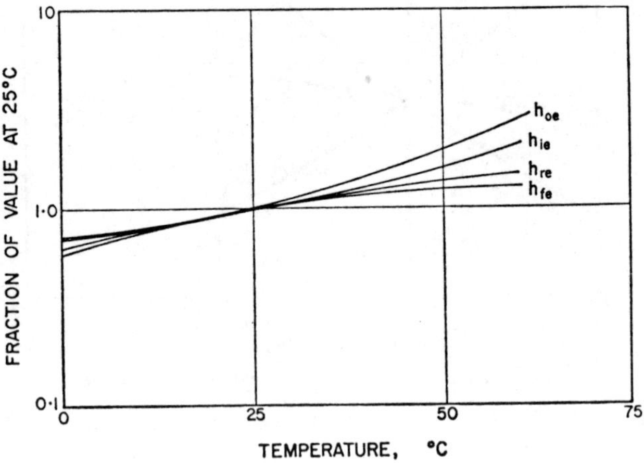

Fig. 6.10.—*Variation of h parameters with temperature.*

current, collector voltage and junction temperature are illustrated in Figs. 6.8, 6.9 and 6.10 respectively. It will be seen that the variation is most severe in the effect of collector current on

h_{ie} and h_{oe}. The change of h_{ie} is due to the fact that the emitter resistance decreases with increase of emitter current, and the change of h_{oe} is due to the decrease of collector resistance with increase of collector current.

In viewing these variations, it should be borne in mind that there are considerable variations between individual transistors of a given type due to the tolerance necessary in the manufacturing processes. Again, the manufacturer normally indicates the order of spread on the parameter values, and any design must take account of the limiting values.

The use of equivalent circuits

The use of an equivalent circuit to investigate the behaviour of a practical low frequency amplifier can best be demonstrated by using the T equivalent circuit. Consider the basic single stage common emitter amplifier shown in Fig. 6.11 (a), with the signal fed to the transistor from a source with an internal resistance R_G and the transistor connected to a load resistance R_L. The full equivalent circuit is shown in Fig. 6.11 (b). This can be analysed for the circuit current gain, the input resistance to the loaded transistor, the output resistance of the transistor connected to the generator and the voltage gain of the amplifier stage.

The circuit current gain (a)

The forward current transfer ratio α is defined as the ratio of collector current to emitter current with the collector short circuited. By evaluating the equivalent circuit it is possible to obtain a general expression for the common emitter circuit current gain for any value of R_L. Consider the full collector loop

$$i_c R_L + i_c r_c + i_e \alpha r_c - i_e r_e = 0$$

also

$$(i_e + i_b + i_c) = 0$$

The circuit current gain $a = i_c/i_b$.

Solving the two equations for i_c/i_b

$$a = \frac{-(r_e - \alpha r_c)}{[r_e + r_c(1-\alpha) + R_L]}$$

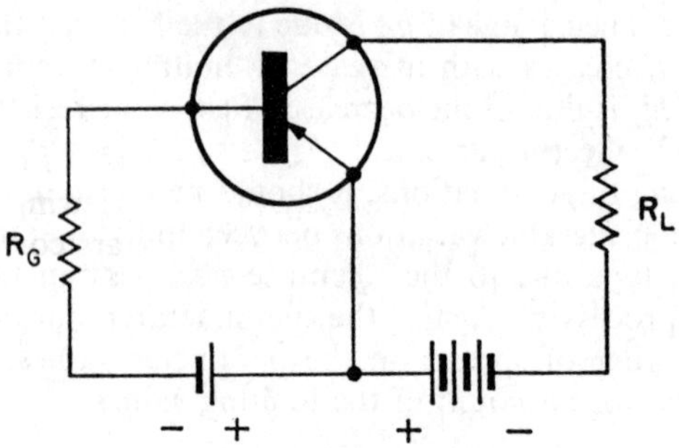

(a)

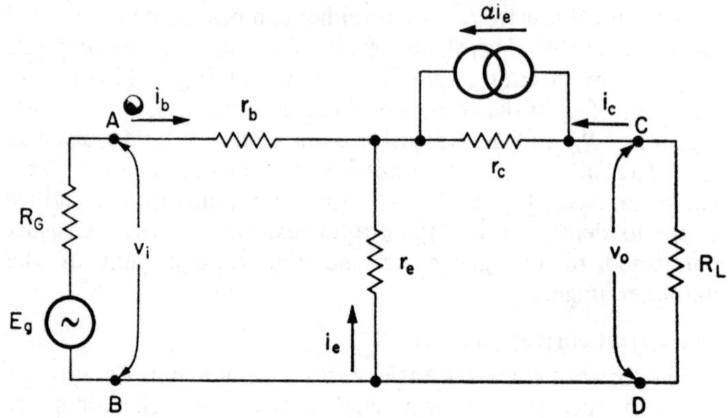

(b)

Fig. 6.11.—*Common emitter amplifier stage.*

r_e is very small compared with r_c and so can be neglected.

Then
$$a = \frac{\alpha r_c}{[r_c(1-\alpha)+R_L]}$$

If $R_L = 0$
$$a = \frac{\alpha}{1-\alpha}$$

which is the short circuit common emitter current gain.

For a practical transistor with $\alpha = 0{\cdot}98$:

If $R_L = 0$ $$a = \frac{0{\cdot}98}{0{\cdot}02} = 49$$

If $r_c = 1\ \text{M}\Omega$ and $R_L = 10\ \text{k}\Omega$

$$a = \frac{0{\cdot}98}{0{\cdot}02 + 0{\cdot}01} = 33$$

Input resistance (R_{in})

Consider the emitter loop, looking into the transistor from terminals AB

$$R_{in} = \frac{v_i}{i_b} = \frac{i_b r_b - i_e r_e}{i_b}$$

also $$(i_e + i_b + i_c) = 0$$

and from above $$\frac{i_c}{i_b} = -\frac{(r_e - \alpha r_c)}{[r_e + r_c(1-\alpha) + R_L]}$$

Combining and solving we obtain

$$R_{in} = r_b + \frac{r_e(r_c + R_L)}{[r_e + r_c(1-\alpha) + R_L]}$$

Usually R_L is small compared with r_c, and r_e is small compared with R_L. With these approximations

$$R_{in} = r_b + \frac{r_e r_c}{[r_c(1-\alpha) + R_L]}$$

For a typical germanium alloy transistor as R_L is increased from zero to 10,000 ohms, R_{in} decreases from 1,900 ohms to 1,400 ohms.

Output resistance (R_{out})

Looking back into the transistor from terminals CD

$$R_{out} = \frac{v_o}{i_c} = \frac{(i_c r_c + \alpha i_e r_c - i_e r_e)}{i_c} \quad \text{also} \quad (i_e + i_b + i_c) = 0.$$

From emitter loop with $E_g = 0$

$$\frac{i_b}{i_c} = \frac{-r_e}{(r_e + r_b + R_G)}$$

Combining and solving we obtain

$$R_{\text{out}} = r_c(1-\alpha) + \frac{r_e(r_b + \alpha r_c + R_G)}{(r_e + r_b + R_G)}$$

Usually r_b and R_G will be small compared with r_c, and r_e is small compared with r_b. With these approximations

$$R_{\text{out}} = r_c(1-\alpha) + \frac{r_e \alpha r_c}{(r_b + R_G)}$$

For a typical germanium alloy transistor, as R_G is increased from zero to 10,000 ohms, R_{out} decreases from 100,000 ohms to 20,000 ohms.

Voltage gain (A_v)

$$\text{We have } A_v = \frac{\text{voltage produced across } R_L}{\text{input voltage}}$$

$$= \frac{-i_c R_L}{v_i}$$

$$= \frac{-i_c R_L}{(i_b r_b - i_e r_e)}$$

Also $$(i_e + i_b + i_c) = 0$$

and $$\frac{i_c}{i_b} = \frac{(r_e - \alpha r_c)}{[r_e + r_c(1-\alpha) + R_L]}$$

Combining and solving we have

$$A_v = \frac{R_L(r_e - \alpha r_c)}{(r_e + r_b)\,[r_e + r_c(1-\alpha) + R_L] + r_e(r_e - \alpha r_c)}$$

Usually r_e and R_L will be small compared with r_c, and approximating

$$A_v = \frac{-R_L}{r_e + r_b(1-\alpha)}$$

With a typical germanium alloy transistor and $R_L = 10\ \text{k}\Omega$, $A_v = 258$.

Alternative circuit arrangements

Similar calculations can be carried out for the other two basic circuit arrangements, common base and common collector, illustrated in Figs. 6.12 and 6.13 respectively.

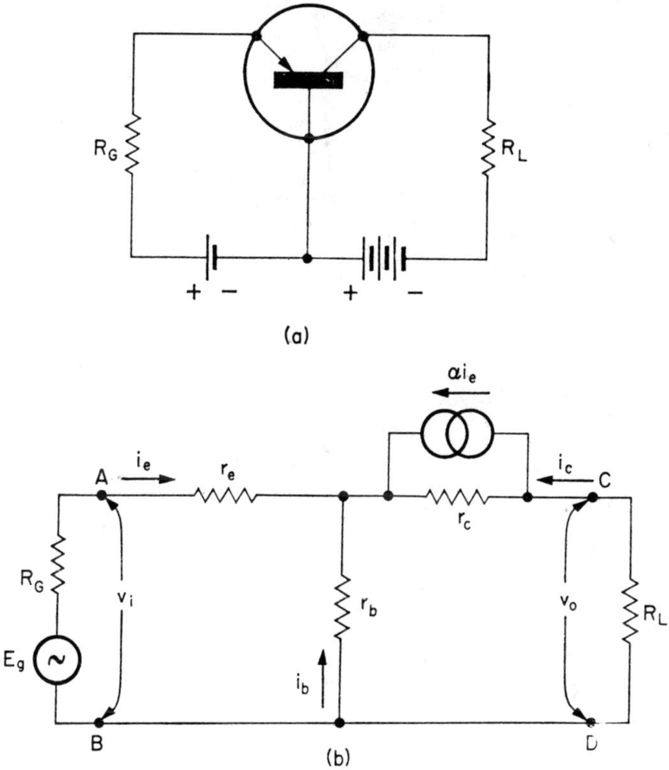

Fig. 6.12.—*Common base amplifier stage.*

Table 6.2 lists the approximate formula for current gain, voltage gain, input resistance and output resistance for the three basic circuit arrangements. Also included in the table are typical values, calculated for a typical germanium alloy transistor with values for $R_L = 10{,}000$ ohms and $R_G = 50$ ohms.

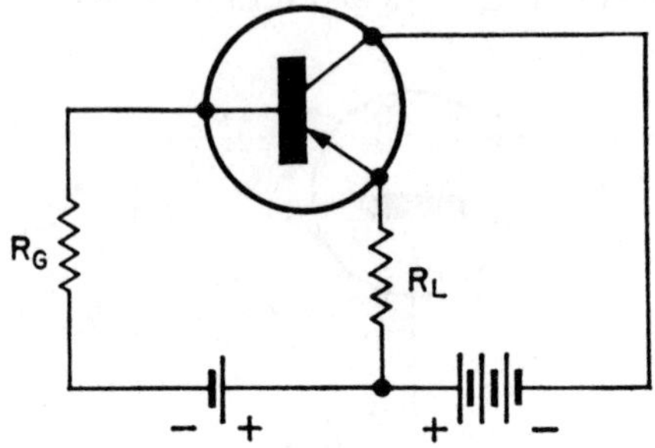

(a)

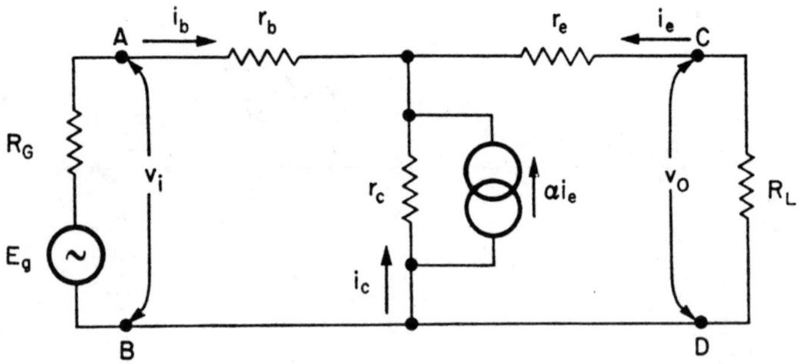

(b)

Fig. 6.13.—*Common collector amplifier stage.*

Perusal of the typical values will give a general picture of relative characteristics of the three basic circuit arrangements. Table 6.3 summarises these characteristics into generalised statements.

Table 6.2. Approximate circuit equations and values

Property		Common base	Common emitter	Common collector
Current gain	Approx. formula	α	$\frac{\alpha}{(1-\alpha)}$	$\frac{1}{(1-\alpha)}$
	Typical values	0·98	49	50
Input resistance	Approx. formula	$r_e+r_b(1-\alpha)$	$r_b+\frac{r_e}{(1-\alpha)}$	$\frac{R_L}{(1-\alpha)}$
	Typical values	38 Ω	1,900 Ω	0·5 MΩ
Output resistance	Approx. formula	r_c	$r_c(1-\alpha)$	$(R_G+r_b)(1-\alpha)$
	Typical values	1 MΩ	20,000 Ω	28 Ω
Voltage gain	Approx. formula	$\frac{\alpha R_L}{r_e+r_b(1-\alpha)}$	$\frac{\alpha R_L}{r_e+r_b(1-\alpha)}$	$\frac{R_L}{r_e+R_L}$
	Typical values	250	250	1

Typical values assuming $r_e = 30\ \Omega$, $r_b = 400\ \Omega$, $r_c = 1\ \text{M}\Omega$, $R_L = 10{,}000\ \Omega$, $R_G = 1{,}000\ \Omega$. (Germanium alloy transistor.)

Table 6.3. Generalised circuit properties

Property	Common base	Common emitter	Common collector
Current gain	Unity	High	High
Input resistance	Low	Medium	High
Output resistance	High	Medium	Low
Voltage gain	High	High	Unity

7

D.C. OPERATING CONDITIONS

THE first consideration in designing a transistor amplifier is the setting of the d.c. operating conditions. The collector d.c. voltage and current must be designed so that the a.c. variations caused by the signal will be within the linear region of the characteristic, in order to avoid distortion to the signal waveform. In some applications, there may be other important aspects which must be considered in deciding the d.c. conditions, for example the noise performance of the circuit. The power rating of the transistor must always be taken into account.

Dynamic considerations—load line

To go further into the considerations involved, it is necessary to look at the operating characteristics of the transistor with its collector load resistance. Consider the circuit shown in Fig. 7.1 (a) and the common emitter collector characteristics of the transistor shown in Fig. 7.1 (b). With the base current at zero, the collector current will be small—the reverse leakage current of the collector junction, and so the voltage between the collector and emitter of the transistor V_{CE} will be almost the full battery voltage of 12 volts. As the base current is increased, the collector current increases, the voltage across the load resistance increases, and the voltage across the transistor decreases. Eventually, when the base current reaches just over 40 μA, the collector current will have increased to such a level that the voltage across R_L approaches the battery voltage, and the voltage across the transistor V_{CE} is almost zero. V_{CE} will never fall quite to zero because there is always a small resistance present in the transistor. Increasing the base current beyond this level will not result in any further increase of collector current, the transistor is said to be saturated, and the collector to emitter voltage in this state is called $V_{CE(sat)}$.

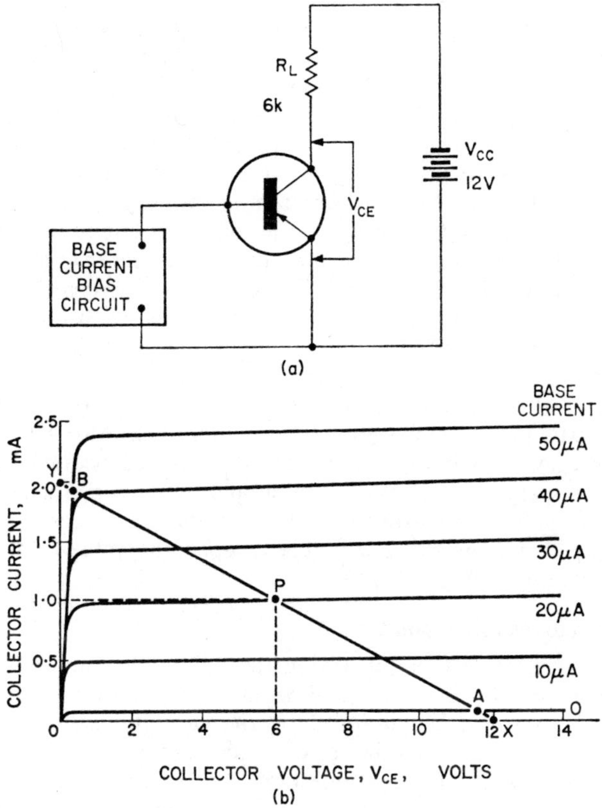

Fig. 7.1.—*Transistor bias considerations.*

This condition corresponds to the point B in Fig. 7.1 (b). By adjusting the d.c. base current between 0 and 40 μA, the transistor operating point can be set to any point along the line AB.

The full line XY is called the *load line* and is constructed on any collector characteristic by starting at the supply voltage, V_{CC}, on the voltage axis and joining to a current on the current axis equal to V_{CC}/R_L. Using the values in the example quoted the load line joins 12 volts to 2 mA.

Operating point

For most general applications the best choice of operating point is mid way along the load line—for example at the point P in Fig. 7.1 (b). In this example, to set the collector current at 1 mA, the base current must be adjusted to 20 μA since the current gain is approximately 50. Now, if an a.c. signal is applied to the base, to cause variation of the base current, an amplified variation will appear on the collector current. It will be seen that for linear amplification the base current variation can be increased to the order of $\pm 20\,\mu$A peak, giving a collector current variation of nearly $\pm 1{\cdot}0$ mA before distortion of the output waveform occurs due to running into the "saturated" condition at the high collector current end, or into "cut-off" at the other end. Any change in the operating point P will result in distortion setting in at a lower level. Also, it was seen in Chapter 6 that the input and output impedances of the transistor vary with collector current, and so the performance of the amplifier stage with regard to such points as signal gain and impedance matching will depend upon the value of the d.c. operating collector current.

Setting the operating point

The simplest method of setting the operating point is to arrange that a d.c. base current flows into the transistor so that when this current is multiplied by the current gain of the transistor the required value of collector current is obtained. With the common emitter connection, this can be arranged by connecting a resistor between the collector voltage supply line and the base of the transistor, as shown in Fig. 7.2. Since the emitter is forward biased, the voltage across the emitter–base junction of the transistor is small, usually less than one volt, and so the required value of the resistor R_b is approximately equal to V_{CC}/I_B. The value of the collector current I_C resulting from the base current I_B will be given by the relation

$$I_C = I_{CBO} + h_{fe}(I_B + I_{CBO}) \qquad (7.1)$$

or transposing

$$I_C = h_{fe} I_B + I_{CBO}(1 + h_{fe}) \qquad (7.2)$$

Examination of equation (7.2) indicates the limitations of this simple method of setting the operating point. First, with any given type of transistor, the specified spread of current gain h_{fe} between different transistors is usually between two and

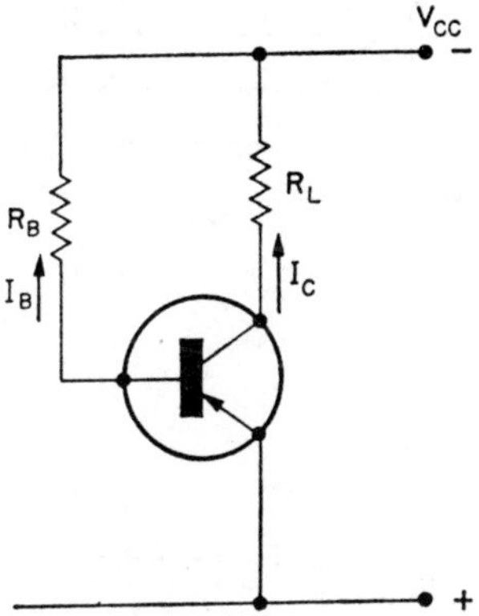

Fig. 7.2.—*Base current bias.*

three to one, and so if I_B is chosen to give the correct value of I_C for one transistor, changing the transistor may result in quite a different value of I_C, and the operating point will be different for each transistor. Secondly, it has been shown in Chapter 6 how the collector leakage current I_{CBO} increases exponentially with temperature. Since, as shown in equation 7.2, I_{CBO} is multiplied by a factor $(1+h_{fe})$ the value of collector current I_C is very temperature dependent. For a typical germanium alloy transistor, the collector leakage I_{CBO} doubles for every 9°C rise in temperature, and although it is only about 1 μA at 20°C it will have increased to about 20 μA at 55°C. If h_{fe} is 50, the component of collector current due to this leakage current will increase from 0·05 mA at 20°C to 1·02 mA at 55°C. Thus, if the operating point had been set to be 1 mA at 20°C, at 55°C it will have increased to 2 mA, and with the values given in Fig. 7.2 the transistor will be just at the saturation point. The undistorted signal output from the amplifier stage will become lower as the temperature increases from

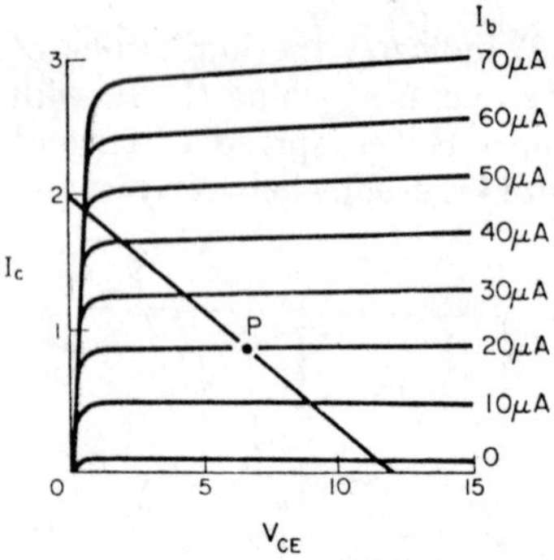

(a) Gain 40 temperature 20°C

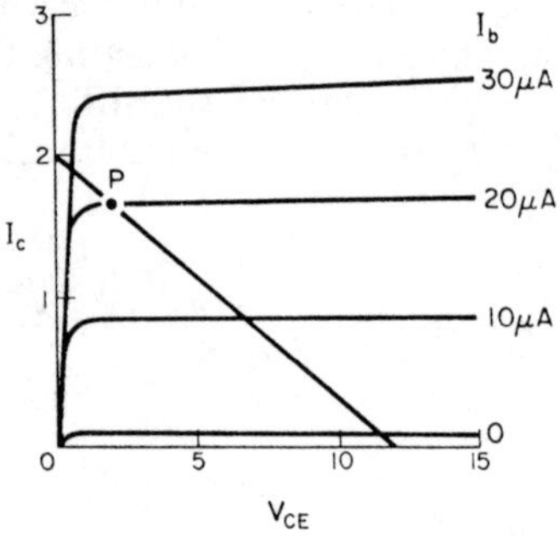

(b) Gain 80 temperature 20°C

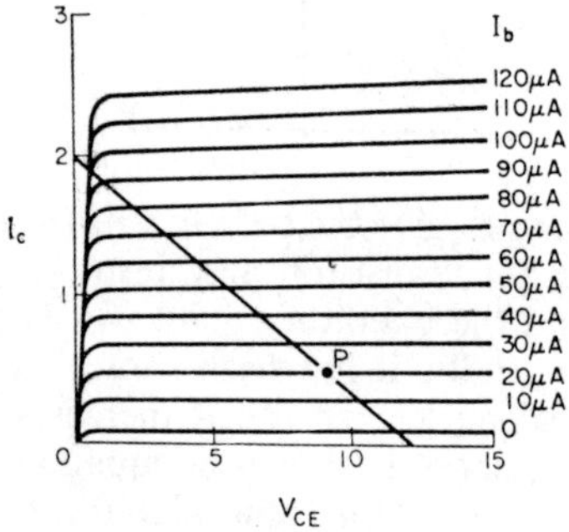

(c) Gain 20 temperature 20°C

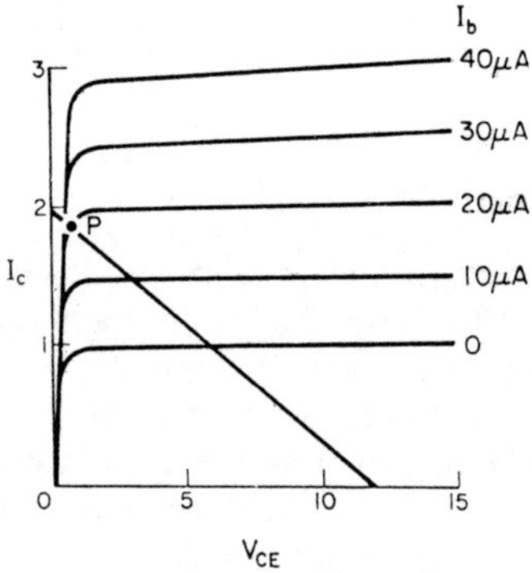

(d) Gain 40 temperature 55°C

Fig. 7.3.—*Variation of operating point.*

20°C, until at 55°C it falls to zero. The change of operating point of an unstabilised base current biased stage due to changes of transistor current gain and temperature are illustrated by the collector characteristics shown in Fig. 7.3.

Bias stabilisation

It can be seen from the previous section that for stable operation of a common emitter transistor amplifier stage special arrangements must be made to ensure that the operating point does not change materially with temperature or when different transistors of the same type are substituted. This subject is usually referred to as *bias stabilisation.* The question

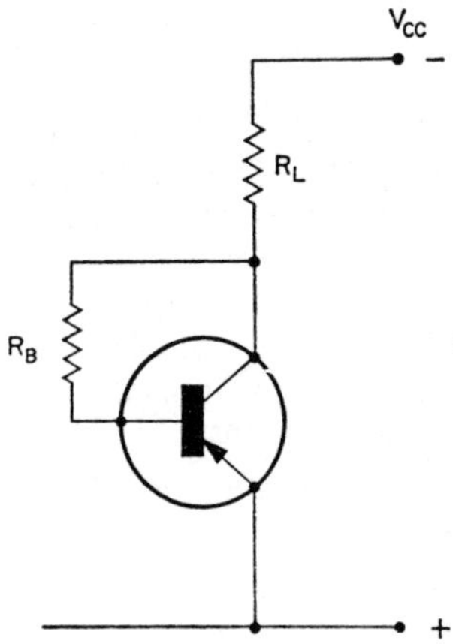

Fig. 7.4.—*Feedback base bias current.*

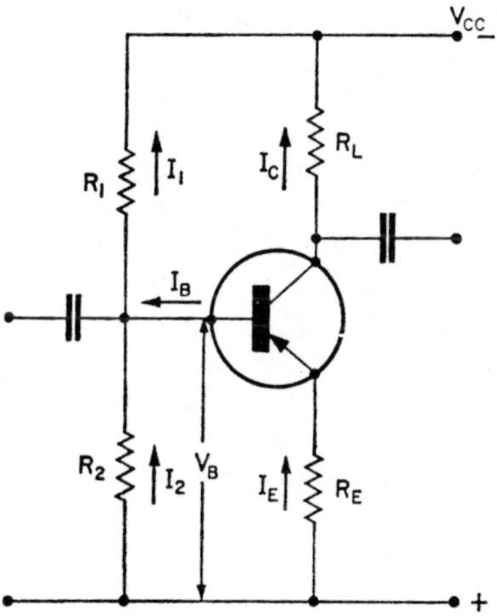

Fig. 7.5.—*Base voltage bias.*

of special stabilisation does not arise with the common base circuit, since changes in common base current transfer ratio, α, are small, and the only leakage current component of the collector current is I_{CBO}. Although this increases with temperature it is so small as to be negligible.

A simple method of obtaining some improvement with the common emitter circuit is to feed the base bias resistor from the collector as shown in Fig. 7.4. If the collector current tends to increase for any reason then the collector voltage will fall and so the base current will fall, tending to cause a reduction in the collector current, thus opposing the initial change. This method is only partially effective, however, as the leakage current with zero base current is unaffected and will still increase with temperature. Also when different transistors are inserted some change of collector current will still result as the stabilisation is not complete. A further limitation is that negative feedback occurs with an a.c. signal, thus causing a reduction in the effective gain of the stage. This feedback can be reduced by splitting the bias resistor and by-passing the centre point to earth with a capacitor.

Base voltage bias

The most widely used method of bias stabilisation is called *base voltage bias* and is illustrated in Fig. 7.5. The bias voltage applied to the base is obtained from a low resistance potentiometer R_1 and R_2 connected across the supply voltage, and a resistance R_E is connected in series with the emitter. The voltage V_B appearing across R_2 will establish the bias condition across the emitter to base junction and will result in an emitter current through the resistance R_E such that the voltage across R_E is equal to V_B less the voltage dropped across the the emitter–base junction of the transistor, i.e.

$$V_B = I_E R_E + V_{EB} \tag{7.3}$$

or

$$I_E = \frac{V_B - V_{EB}}{R_E} \tag{7.4}$$

If V_B is made large compared with V_{EB}, the emitter current I_E will depend only on V_B and R_E. Thus if V_B can be held constant, I_E, and so I_C, will be independent of temperature and transistor gain. Any increase in collector to base leakage current I_{CBO} due to increase in temperature will flow from the base of the transistor down through R_2. If the value of R_2 is sufficiently low this will not cause a significant change in V_B.

The performance of the circuit in stabilising the bias point is defined by a factor S, which equals the change in total collector current divided by the change in collector to base leakage current, i.e.

$$S = \frac{\delta I_C}{\delta I_{CBO}} \tag{7.5}$$

Ideally this would be unity as in the case of the common base circuit. In the unstabilised common emitter arrangement of

Fig. 7.2, S will equal the common emitter current gain, typically in the range 25 to 100. Practical values of S for most application work are in the region of 3–6.

The circuit of Fig. 7.5 can be analysed to obtain an expression for the value of S to assist in circuit design. Several relations can be written down

$$I_B = I_1 - I_2 \tag{7.6}$$

$$I_C = I_E - I_B \tag{7.7}$$

$$V_{CC} = I_C R_L + I_E R_E + V_{CE} \tag{7.8}$$

$$I_2 R_2 \simeq I_E R_E \tag{7.9}$$

$$I_1 R_1 = I_C R_L + V_{CE} \tag{7.10}$$

$$I_C = I_{CBO} + \alpha I_E \tag{7.11}$$

Evaluating we find

$$S = \frac{1 + R_E\left(\dfrac{1}{R_1} + \dfrac{1}{R_2}\right)}{(1-\alpha) + R_E\left(\dfrac{1}{R_1} + \dfrac{1}{R_2}\right)} \tag{7.12}$$

Examination of this relation indicates that to make S low we must make R_E high and R_1 and R_2 low. The extent to which R_E can be increased will depend upon how much of the supply voltage is available to be dropped across it. One limit on how low R_1 and R_2 can be reduced is the drain current allowable from the power supply line. Since the input impedance of a common emitter stage is about 1,000 ohms, R_2 should not be so low as to reduce this materially, since it is effectively in parallel.

Practical values for a typical silicon alloy transistor stage are given in Fig. 7.6.

V_B will equal $12 \times 4{\cdot}7/26{\cdot}7 = 2{\cdot}1$ V

V_{EB} will be approximately $0{\cdot}5$ V

Thus $$I_E = \frac{1{\cdot}6}{1{,}000} = 1{\cdot}6 \text{ mA}$$

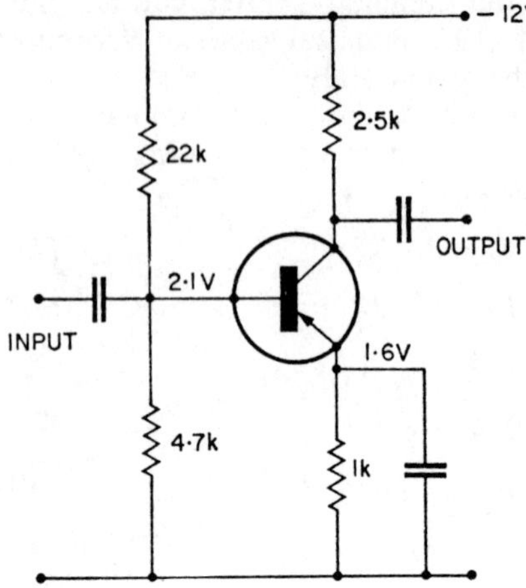

Fig. 7.6.—*Typical common emitter amplifier stage.*

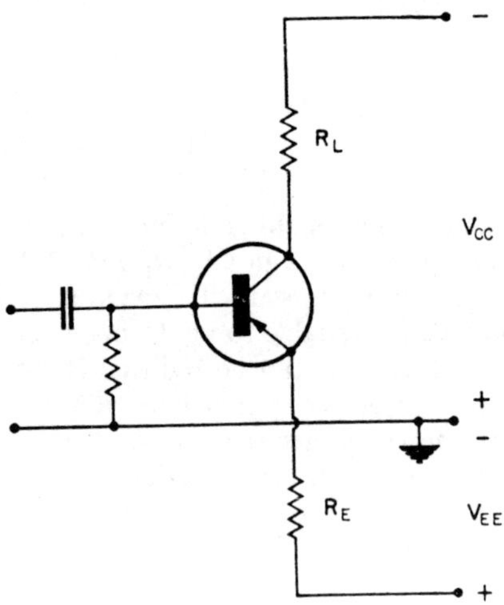

Fig. 7.7.—*Emitter bias.*

If I_{CBO} increases from 1 μA at 20°C to 20 μA at 55°C, the increase in V_B due to this will be $4{,}700 \times 20/10^{-6} = 0{\cdot}1$ V which is negligible compared with 2·1 volt.

Evaluating S from equation (7.12) we have $S = 4{\cdot}5$.

To avoid loss of gain of an a.c. signal the emitter resistance R_E is by-passed by a large capacitor, so that the input signal appears directly between base and emitter.

Emitter bias

A further method convenient in some applications is called emitter bias, in which the base is at earth potential and the emitter is returned to a potential below earth, as shown in Fig. 7.7. In this circuit, the current drain through the potential divider is avoided. The emitter current is approximately equal to $(V_{EE}-V_{EB})/R_E$. For good stabilisation the voltage V_{EE} must be high compared with the emitter–base voltage V_{EB} of the transistor.

8
LOW LEVEL LOW FREQUENCY AMPLIFIERS

In low level amplifiers, it is assumed that the signal level is small, such that the signal "swing" is only over a small region of the load line around the d.c. operating point. The a.c. characteristics of the transistor can be assumed to be constant as long as bias stabilisation is used. The three basic arrangements of a single stage transistor amplifier have already been discussed and the versatility of the transistor is illustrated by the range of input and output resistances possible from the various configurations.

A low level amplifier will usually be required to accept a small signal—possibly down to the order of micro-volts—and amplify to a level high enough to feed the driver stage of a power amplifier or to operate a relay or indicating meter. Only on rare occasions will a single stage be sufficient, two or three stages being more usual. General points for consideration in the design of an amplifier are circuit configuration, number of stages, bias stabilisation, interstage coupling, gain control, frequency response, negative feedback and first stage signal-to-noise performance. These various points will be discussed in turn.

Circuit configuration

The *common emitter circuit* is the most convenient arrangement for low frequency amplifier circuits. As will be shown later, its high current gain makes possible the use of resistance capacity coupling. Although the input resistance is normally of the order of one to two thousand ohms, it can readily be increased by inserting an unby-passed resistor in the emitter circuit. Referring back to the relation for input resistance (Chapter 6) we have

$$R_{\text{in}} = r_b + \frac{r_e}{(1-\alpha)}$$

For the typical small signal germanium alloy transistor

$$R_{\text{in}} = 400 + \frac{30}{0{\cdot}02} = 1{,}900 \text{ ohms}$$

With a resistor R_E connected in series with the emitter

$$R_{\text{in}} = r_b + \frac{(r_e + R_E)}{(1-\alpha)}$$

If R_E is made 220 ohms,

$$R_{\text{in}} = 400 + \frac{250}{0{\cdot}02} = 12{,}900 \text{ ohms}$$

There will be a loss of signal across R_E and, since it is common to both input and output circuits, it will also introduce negative feedback, thus reducing the stage gain.

The input resistance can also be increased by including a resistor R_B in series with the input to the base. Then

$$R_{\text{in}} = (r_b + R_B) + \frac{r_e}{(1-\alpha)}$$

Again there will be a loss of signal across R_B.

A third method of obtaining some control of input resistance is by the choice of the operating emitter current. Reducing the emitter current will increase r_e and so will increase the input resistance.

Another important feature of the common emitter circuit is that it results in a phase reversal of the signal. Consider an npn stage, if the base voltage is made more positive, the base current increases, the collector current increases and the collector voltage falls, so that it becomes more negative. This phase reversal facilitates the use of negative feedback.

The *common base circuit* gives voltage gain only. The current gain is always less than unity and so resistance capacity coupling is not possible. In order to obtain a power gain with coupled common base stages it is necessary to use a step down transformer to match the high output resistance of one stage to the very low input resistance of the next stage. The main use of the common base stage in low frequency amplifiers is where a very low input resistance, of the order of ten or twenty ohms, is required.

The *common collector circuit*, more often called the emitter-follower circuit, is useful to obtain a very high input resistance with a low output resistance. The high input resistance is obtained due to the fact that the load resistance is in series with the emitter, as discussed above. The current gain is high, approximately the same as for the common emitter, but the voltage gain cannot exceed unity.

Number of stages

It is necessary at the start of any amplifier design to estimate the number of stages required. The required overall gain can be deduced from the input available and the output required. Then, consideration of the expected gains from individual stages together with the losses in any interstage coupling networks will allow a first estimate of the number of stages required. With low level transistor amplifiers it is usually best to consider the stages as current amplifiers, and to evaluate the current gain requirements. In some applications it may be convenient to include a stage mainly for some manipulation other than signal amplification, such as impedance transformation or frequency response control. Also with many amplifiers some negative feedback will be included, and in this case the gain requirements of the amplifier without feedback must be greater, depending upon the amount of feedback.

Interstage coupling

The a.c. output signal from a common emitter stage can be coupled to the input of the next stage either by means of a transformer or by a resistance capacitance combination. Maximum power gain will be obtained with a transformer having a suitable primary to secondary step down ratio to match the output resistance of the first stage to the input resistance of the second stage. The turns step down ratio will typically be between five and ten to one. Transformers, however, are relatively expensive components and are also susceptible to magnetic pick-up. Accordingly they are rarely used in low level transistor amplifiers.

With resistance capacitance coupling the input resistance of the second stage effectively becomes the load resistance of the first stage, and so the input resistance and load resistance of the first stage are the same. Thus the voltage gain will equal the current gain and the power gain will be equal to the current gain squared. If the current gain is 50, the power gain will be 2,500, i.e. 35 db.

The effective a.c. load on the transistor consists of several components in parallel—the load resistance R_L, the output resistance of the transistor, the base bias resistors R_1 and R_2 and the input resistance of the next transistor. The only useful part of this load is the input resistance of the next transistor—the other components all represent loss, and when evaluating the overall gain of a transistor amplifier it is necessary to take these losses into account. This is illustrated later in the chapter when a typical evaluation is presented.

The value of coupling capacitor depends upon the required low frequency response of the amplifier. As the signal frequency is decreased, a point will be reached where the amplifier gain starts to decrease due to the increasing impedance of the coupling capacitor. The gain will have fallen by 30 per cent (3 db) at the frequency at which the reactance of the coupling capacitor equals the total resistance in series with it. This series resistance is equal to the load resistance of the previous stage plus the input resistance of the next stage.

Referring to Fig. 8.1

$$C_1 = \frac{1}{2\pi f(R_L + R_{\text{in}})}$$

With $R_L = 2{\cdot}5$ kΩ and $R_{\text{in}} = 1{\cdot}5$ kΩ, for a 3 db loss at 25 c/s a coupling capacitor of $1{\cdot}5$ μF will be required.

It must be remembered that the base bias resistors R_1 and R_2 are both effectively in parallel with the input resistance of the transistor and so must be taken into account.

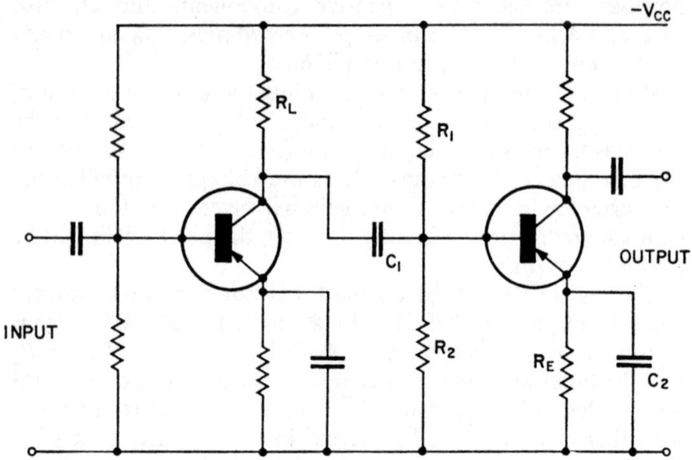

Fig. 8.1.—*Resistance-capacitance coupling.*

Emitter resistance decoupling capacitor

The value of the emitter resistance decoupling capacitor C_2 also depends upon the required low frequency response. The gain of the amplifier will fall by 3 db at the frequency at which the reactance of capacitor C_2 equals the emitter resistance R_E shunted by the resistance presented by the transistor. This latter is approximately the output resistance of the transistor in the common collector arrangement and equals

$$(R_G+r_b)(1-\alpha).$$

R_G will be R_L in parallel with R_1 and R_2, and, with the values assumed earlier in the section, $R_G = 1{,}670$ ohms, giving a transistor output resistance of approximately 40 ohms.

The required capacitance for a 3 db fall in gain at 25 c/s will be 150 μF.

Approximate analysis of a two-stage common emitter amplifier

Consider the amplifier circuit shown in Fig. 8.2 using type 2N338 small signal silicon transistors. The collector supply voltage is 12 volts, the operating point of the first transistor $V_{CE} = 5$ volts, $I_E = 1$ mA and of the second transistor $V_{CE} = 5$ volts, $I_E = 6$ mA. The operating points are set by fixing the values of the bias, load and emitter resistors as

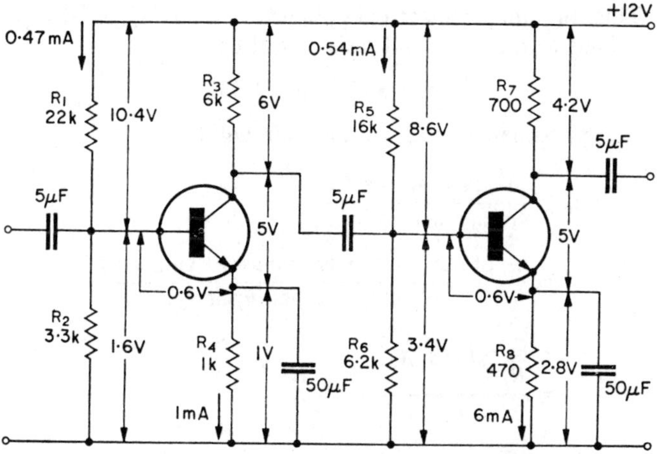

Fig. 8.2.—*Analysis of a two-stage audio amplifier.*

shown. The various direct voltages and currents will be as indicated. From the manufacturer's data, the common emitter current gain of the transistor at 1 mA is 49, and when corrected for 6 mA is 66. Also from the manufacturer's data the input resistance of the transistor at 6 mA is 700 ohms and at 1 mA 2,100 ohms. For the following analysis, the effect of the transistor output resistance is neglected.

Considering the output current flowing through the 700 ohms collector load, the current gain of the second stage will be 66.

The current gain of the first transistor will be 49, but this will be reduced by the loss in the interstage components R_3, R_5 and R_6. The effective resistance of these three in parallel is 2,500 ohms. This resistance in parallel, with the input resistance of 700 ohms, will decrease the first stage gain to $49 \times 2{,}500/3{,}200 = 38{\cdot}2$.

Similarly the input circuit to the first transistor will cause a signal current loss. Resistors R_1 and R_2 in parallel (2,800 ohms) will appear across the input resistance (2,100 ohms) of the first transistor. The proportion of current entering the transistor will be 0·57.

Thus the approximate overall current gain, A_i, from input terminals to the final load resistor will be

$$0{\cdot}57 \times 38{\cdot}2 \times 66 = 1{,}437.$$

The approximate overall voltage gain will be

$$A_v = \frac{\text{voltage across } R_7}{\text{voltage across input}}$$

$$= \frac{\text{input current} \times \text{overall current gain} \times R_7}{\text{voltage across input}}$$

$$= \frac{\frac{e_i}{R_i} 1{,}437 \times 700}{e_i},$$

where R_i = effective resistance across input = 1,140 ohms,

$$= \frac{1{,}437 \times 700}{1{,}140}$$

$$= 880$$

Gain control

The gain of a transistor amplifier can be adjusted by making the coupling network between two stages a variable attenuator. The attenuator must be designed so that it has negligible effect on the frequency response of the network or the d.c. operating point of either of the transistors. Fig. 8.3 shows two suitable methods of control. That shown at A is probably the better and is most widely used. It has no effect on the bias setting and has only a slight effect on the low frequency response. The control shown at B has no effect on the low frequency response, but will alter the d.c. voltage across the first transistor.

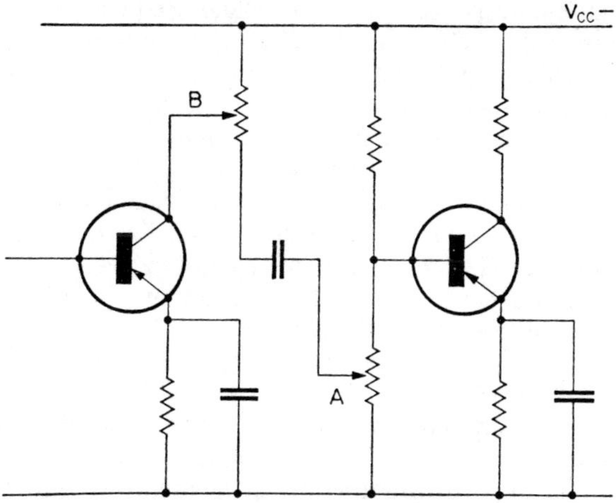

Fig. 8.3.—*Gain control.*

Frequency response control

In audio amplifiers control of frequency response is usually referred to as "tone control" and is used to accentuate or

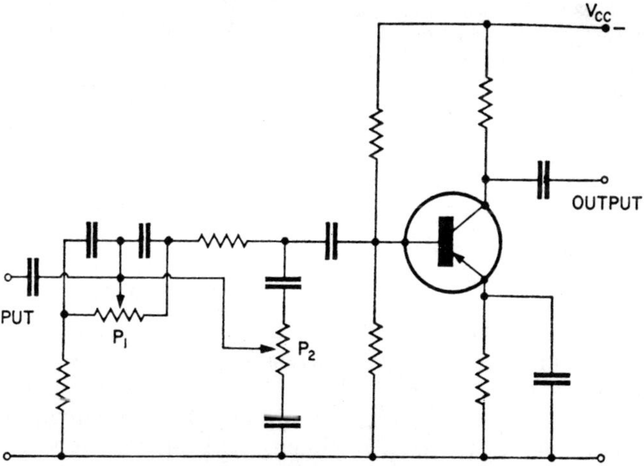

Fig. 8.4.—*Tone control.*

attenuate the response at the low or high frequency end of the frequency band. The control can be carried out by incorporating frequency dependent circuits in the coupling network between two stages. Increasing the effective capacitance in parallel with the load resistance of a transistor will reduce the high frequency response, and by varying the effective capacitance in the coupling circuit the low frequency response can be controlled. A typical arrangement is shown in Fig. 8.4. Potentiometer P_1 controls the low frequency response and P_2 the high frequency response. Further examples will be seen in the circuits at the end of the chapter.

Negative feedback

The gain of an amplifier can be stabilised and the internally generated distortion reduced by introducing negative feedback. This can be applied from the output to input of a single stage,

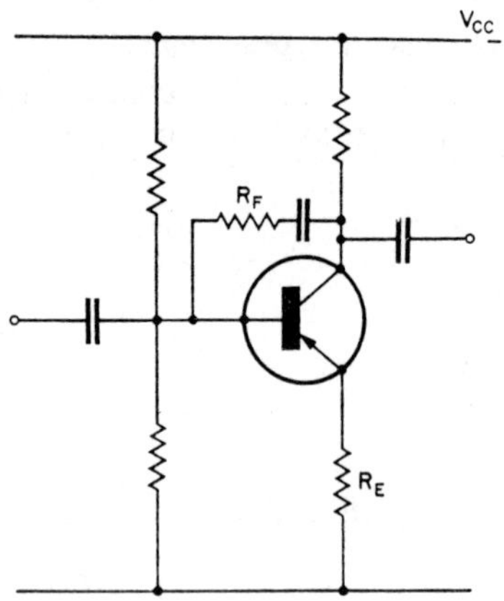

Fig. 8.5.—*Application of negative feedback over a single stage.*

or alternatively over several stages. With single-stage common emitter amplifiers, negative feedback can be obtained in two ways—by feeding back from the collector to base as shown (Fig. 8.5) via R_F, or by including an unby-passed resistor R_E in the emitter circuit. The latter is very useful and is commonly used. It has already been seen that such

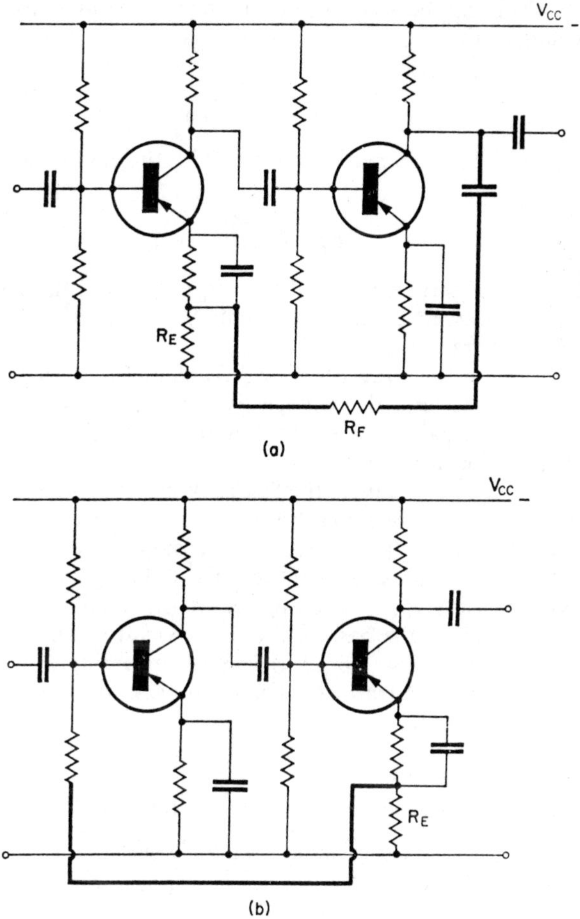

Fig. 8.6.—*Shunt and series feedback. Heavy lines indicate feedback paths*

a resistor increases the input resistance to the transistor. This reduces distortion caused by the non-linear characteristic of the base emitter "diode". At the same time some loss of gain results. This can be seen by considering approximate formula for voltage gain.

$$A_v = \frac{\alpha R_L}{[r_e + r_b(1-\alpha)]}$$

If a resistor R_E is connected in series with the emitter we have

$$A_v = \frac{\alpha R_L}{[R_E + r_e + r_b(1-\alpha)]}$$

If R_E is made large compared with $[r_e + r_b(1-\alpha)]$ the voltage gain is reduced and becomes essentially independent of the transistor characteristics.

For the typical transistor—see Table 6.2—$[r_e + r_b(1-\alpha)]$ equals 40 ohms, and the voltage gain is 250. If R_E is made 400 ohms, the gain falls to 22, but becomes relatively independent of the transistor.

In multi-stage amplifiers the feedback may be introduced in series with the source signal (series or current feedback) or in parallel with the input signal (shunt or voltage feedback). Series feedback is illustrated in Fig. 8.6 (b), the signal developed across a resistor R_E in the emitter of the second transistor being fed back to the base of the first transistor. Shunt feedback is shown in Fig. 8.6 (a), the signal appearing at the collector of the second transistor being fed back to the emitter of the first transistor.

Low noise considerations

The noise figure of an amplifier is defined by the equation

$$\text{N.F.} = \frac{\text{input signal-to-noise power ratio}}{\text{output signal-to-noise power ratio}}$$

An ideal amplifier would amplify the incoming signal and noise without changing the signal-to-noise ratio, that is, the noise figure would be unity (i.e. 0 db). In practice, some extra noise is generated in the transistor and so the noise figure of a transistor amplifier is always greater than unity. The nearer the noise figure is to unity, the better is the amplifier.

If an amplifier is required to amplify signals with amplitudes of the order of microvolts, special attention must be paid to the design from the low noise aspect. The noise figure of a transistor amplifier is dependent upon the value of emitter current

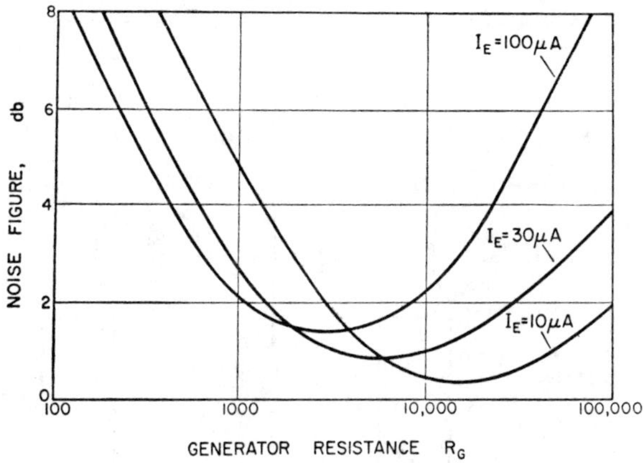

Fig. 8.7.—*Relation between noise figure and generator resistance.*

and the source resistance. As the emitter current is reduced, the noise figure falls, but eventually it starts to rise again as the current gain decreases at very low currents. In general, the emitter current for best low noise operation will be between 10 μA and 200 μA. For germanium transistors it will be towards the top of this range, and with silicon towards the lower values. For a given value of emitter current there will be a certain value of source resistance for minimum noise figure. As the emitter current is decreased, the optimum value of source resistance increases. These points are illustrated in Fig. 8.7 which shows the relation between noise figure and source resistance for three different values of emitter current for a silicon planar transistor. In deciding upon the value of emitter current for a particular application, other considerations such as level of leakage current and stability against temperature must be taken into account, and so a compromise may be necessary.

Fig. 8.8 shows a typical low noise amplifier circuit using 2N930 silicon planar transistors. The first stage operates with $I_E = 10\,\mu A$, and is designed to have a high gain so that the signal is amplified to a level well above the effective noise of the second stage. The effective noise level at the input of the amplifier will be of the order of $0{\cdot}5\,\mu V$, and so a signal of $10\,\mu V$ will give a signal to noise ratio of 20. The second stage has a collector current of $100\,\mu A$.

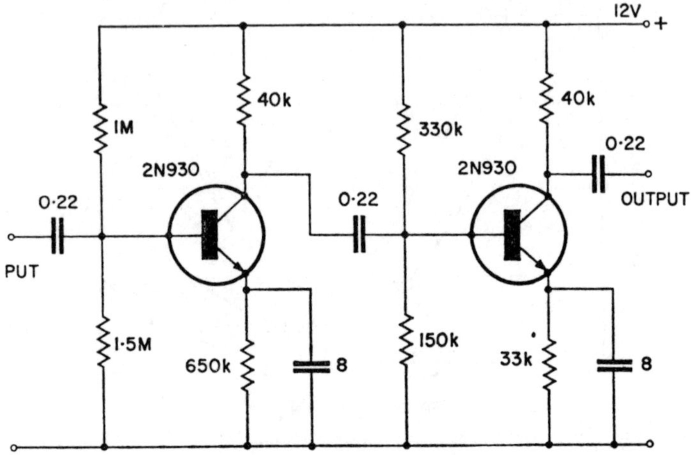

Fig. 8.8.—*Low noise amplifier.*

PRACTICAL EXAMPLES OF LOW FREQUENCY AMPLIFIERS

To illustrate the various factors discussed in this chapter, several practical circuits will now be presented.

Hearing aid

A hearing aid circuit using germanium alloy transistors is shown in Fig. 8.9. The three-stage amplifier operates from a single 1·5 volt battery, and since the unit will not be subjected to very great changes of temperature while in use the d.c. operating point is set by the collector-to-base resistor method. This affords sufficient bias stabilisation, and at the same time introduces a small amount of a.c. negative feedback. A medium-impedance magnetic type microphone—about

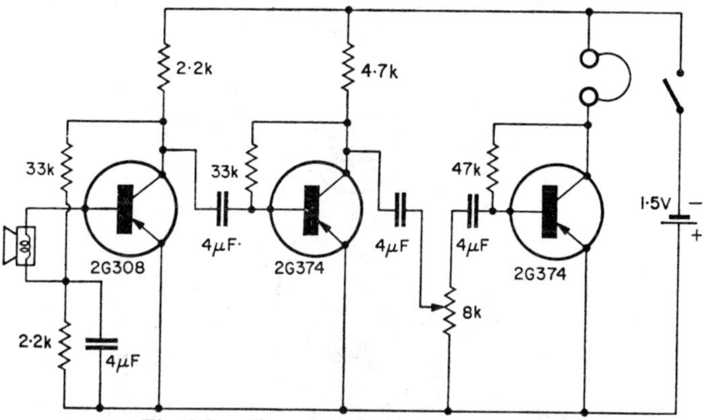

Fig. 8.9.—*Three-stage hearing aid amplifier.*

600 ohms—will match directly into the first transistor. The earphone is connected in the collector circuit of the final transistor.

Electronic stethoscope amplifier

This is similar to a hearing aid amplifier, but has higher sensitivity and a different frequency response. The circuit is shown in Fig. 8.10. Four amplifier stages are used, and the frequency response can be set to either of two conditions. The first, for listening to breathing sounds, has a peak response around 1 kc/s and "rolls off" at a rate of about 6 db per octave. In the second condition, for listening to heart beats and similar low frequency sounds, the capacitor *C* is switched into circuit to cut the medium and high frequency response. The amplifier operates from a 3 volts supply.

Audio preamplifier

A preamplifier circuit that is suitable for amplifying magnetic pick-up signals to a level suitable for feeding a power amplifier is shown in Fig. 8.11. This is a three-stage amplifier using

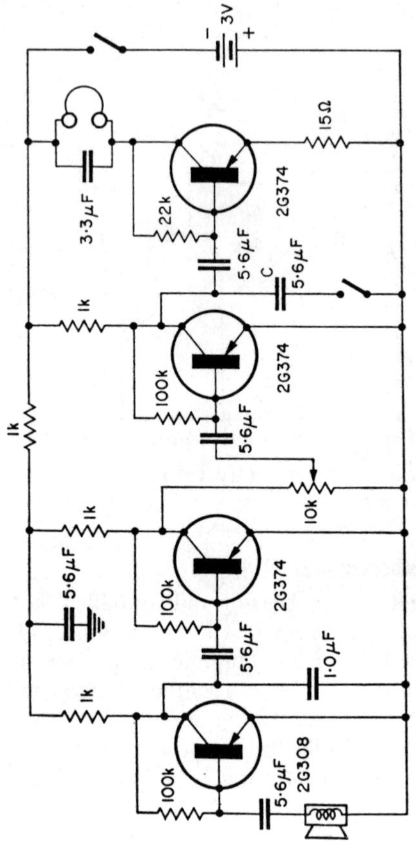

Fig. 8.10.—Electronic stethoscope amplifier.

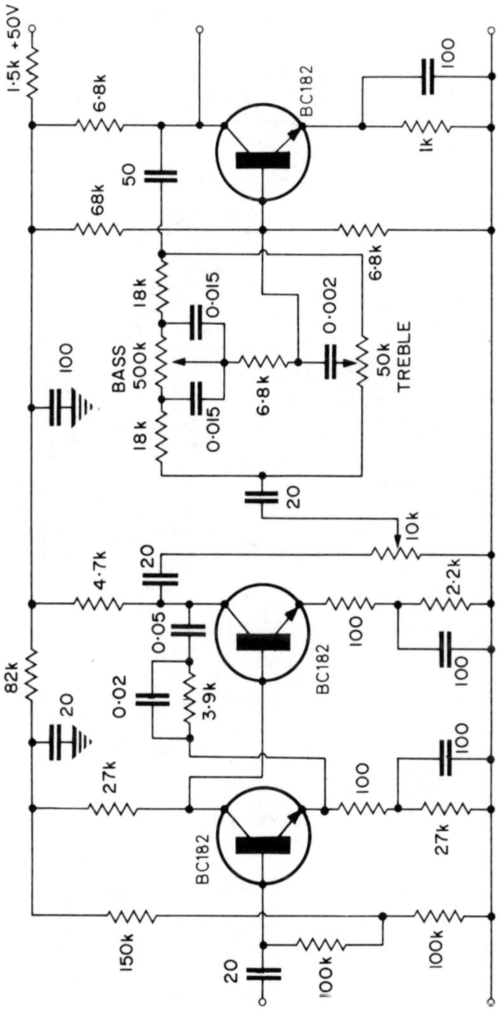

Fig. 8.11.—*Audio preamplifier.*

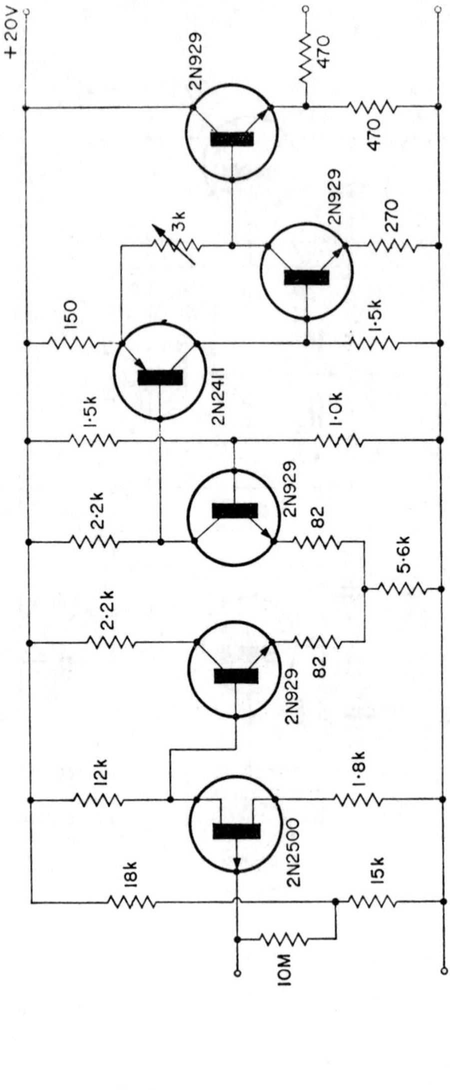

Fig. 8.12.—*Low noise, high input impedance amplifier.*

general purpose silicon planar transistors and incorporating bass and treble tone controls. The first stage is operated at low collector current for low noise operation, and is d.c. coupled to the second stage. The output from the second stage is fed via a volume control to the final stage. Negative feedback is applied from the collector of the second transistor back to the emitter of the first stage, to give the required response to compensate for the recording characteristic, 20 db boost at 20 c/s and 20 db cut at 20 kc/s. The tone control circuit is also in the form of negative feedback between the collector and base of the third transistor. The base tone control gives 17 db boost to 13 db cut at 20 c/s and the treble control 17 db boost to 20 db cut at 20 kc/s.

The output circuit is designed to feed directly into the 20 watt power amplifier described in the next chapter.

Low noise high input impedance amplifier

The use of a silicon field effect transistor to give a high input impedance amplifier with low noise performance is shown in Fig. 8.12. After the first stage, silicon planar transistors are used with d.c. coupling throughout. The first stage has an input resistance of the order of 10 MΩ with an input capacitance of 8 pF. When fed from a source resistance of 1·5 MΩ, the amplifier has a noise figure of 1 db, making it suitable for amplifying signals in the microvolt region. The first stage is followed by an emitter coupled pair of low noise 2N929 npn transistors, the output being direct coupled to a pnp emitter follower which feeds a further npn stage and an emitter follower output. This circuit is further discussed in Chapter 14.

The voltage gain of the amplifier is 100. The frequency response of the amplifier proper is d.c. to 1 Mc/s, but the input capacitance of 8 pF across the source resistance of 1·5 MΩ will limit the response to 10 kc/s. A lower source resistance will widen the response but the noise performance will be slightly worse. The output impedance from the final emitter follower is padded to 500 ohms by the series 470 ohms resistor.

9

HIGH POWER AUDIO AMPLIFIERS

THE important aspect in which a power amplifier differs from the small signal amplifier is that the signal amplitude can extend over the greater part of the dynamic load line of the ouput transistors, and so the transistors will in general be required to operate up to their maximum allowable power levels. An amplifier can be a "power" amplifier irrespective of the actual maximum value of power output, which will depend upon the output transistors used, the circuit design and the input power. Thus we may have a power amplifier using low power transistors delivering an 0·5 watt output, or one using high power transistors delivering 20 watts—the design considerations are similar for both.

The power output from a transistor amplifier will be limited by three main factors associated with the output transistors—the maximum allowable collector dissipation, the maximum collector voltage and the maximum collector current. The allowable power dissipation can be controlled to some extent by the user by suitably designing the thermal properties of the transistor mounting system. The other two factors depend upon the design and manufacture of the transistor and are specified by the manufacturer.

Classification of output stages

For audio power amplifiers, the output transistors can be operated in one of two modes of bias. With class A operation the transistor is biased at the mid point of the load line so that the signal swing is symmetrical about the bias point. With class B operation the bias point is close to cut-off so that upward signal swing increases the collector current, and during the downward swing the collector current remains at zero. Thus with class B operation it is necessary to have two transistors, each biased to cut-off and arranged so that the upward signal swing increases the collector current of one transistor and the downward swing is reversed in phase to increase the collector current of the other transistor. The two collector currents are combined to reform the amplified signal.

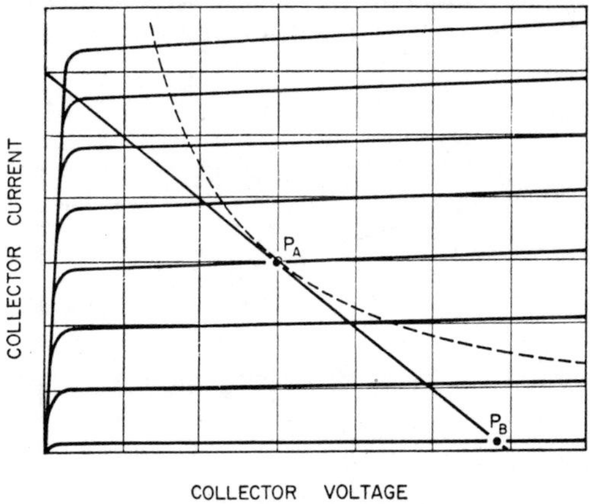

Fig. 9.1.—*Operating points.*

These bias points are illustrated in Fig. 9.1. The maximum power dissipation curve is indicated by the dotted hyperbola. To obtain maximum output in class A, the operating point must be on the hyperbola and be approximately at the mid point of the load line as at P_A. For class B the operating point will be at P_B.

Power transistor characteristics

The maximum power dissipation of a power transistor is limited by the maximum allowable temperature of the collector junction. There are certain power losses in the transistor, and these losses will generate heat which will cause the internal temperature of the transistor to rise. The final temperature reached will depend upon how fast we can arrange for the heat to flow out from the transistor and be transferred to the surroundings. To help this, power transistors are usually clamped on to a fin system with good heat transfer properties.

The maximum collector voltage of a transistor has been discussed earlier and depends largely on the semiconductor material used in the manufacture. In the design of the transistor it is usually necessary to allow some compromise between the maximum voltage and other characteristics such as series resistance, high frequency performance and current amplification.

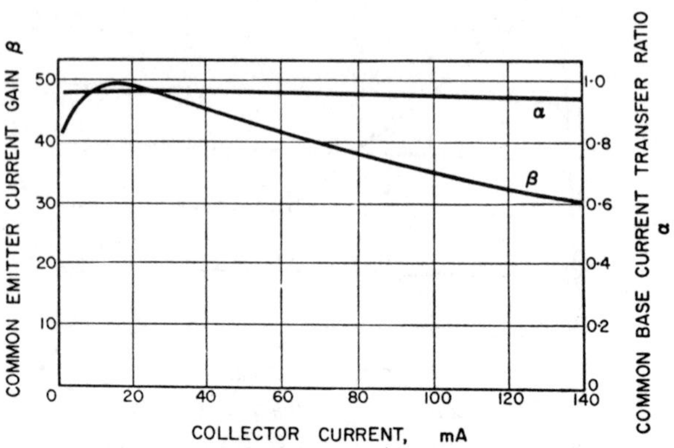

Fig. 9.2.—*Common emitter current gain fall-off.*

The maximum collector current rating is rarely dictated purely by the current carrying capacity of the internal structure. It is usually decided by consideration of the fact that the current amplification, β, falls off as the emitter current is increased, and so the maximum current rating is usually established as the current at which the current gain has fallen by a certain amount. This fall-off in current gain is due to the fact that at high emitter current density the emitter efficiency falls. Although this may result in only a small drop in the common base current gain, α, when this is converted to the common emitter current gain, β, the decrease becomes relatively large. This is illustrated in Fig. 9.2. A fall in α from 0·98 to 0·97 results in a fall in β from 49 to 32. As the emitter current increases, the current gain characteristic becomes non-linear and so distortion occurs in the ouput signal at high level. Thus it is necessary to limit the emitter current in order to limit the distortion level. The effect can be reduced by making the emitter and collector larger in area, so as to reduce the current density for a given current, but this will result in increased capacitance and poorer high frequency response.

Class A output stage

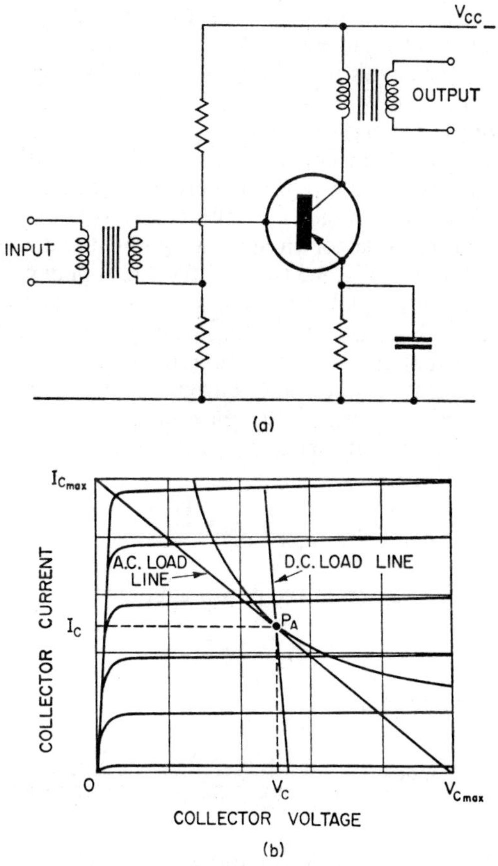

Fig. 9.3.—*Basic Class A stage and characteristics.*

A class A output stage can be used when relatively low output is required and when a high quiescent current drain is allowable. The main criterion for the design of any power output stage is usually that of ensuring the maximum possible output power. An output coupling transformer is generally used to match the output to the actual load as in the basic class A stage illustrated in Fig. 9.3 (a). For maximum power output, the operating point should be on the maximum power hyperbola as shown at P_A in Fig. 9.3 (b) and the a.c. load resistance in the collector circuit must be arranged so that the maximum rating of the transistor cannot be exceeded when maximum signal swing is experienced. The a.c. load line is drawn from V_{Cmax}, the maximum collector rating of the transistor, tangential to the maximum power hyperbola, the inverse slope of the line being equal to the value of the a.c. load. Since the d.c. resistance in the collector circuit is very low, the d.c. load line will be as shown in Fig. 9.3 (b), and the d.c. supply voltage will be approximately equal to the d.c. collector voltage V_C which is one half of the maximum voltage rating of the transistor.

Assuming that the transistor characteristics are linear, and that the output voltage and current swing from zero to maximum and vice versa, the maximum power output will equal the product of the r.m.s. voltage and r.m.s. current swing, i.e.

$$P_o = \frac{V_C}{\sqrt{2}} \times \frac{I_C}{\sqrt{2}} = \frac{V_C I_C}{2}$$

The power supplied from the power supply to the transistor

$$P_S = V_C I_C$$

Thus, neglecting losses,

$$\text{the efficiency} = \frac{\text{output power into the load}}{\text{total input power from the supply}}$$

$$= \frac{V_C I_C/2}{V_C I_C} \times 100 = 50 \text{ per cent}$$

This is the maximum theoretical efficiency possible with class A operation. In practice, efficiencies of the order of 48 per cent can be obtained.

With the signal swing at zero, the power input from the supply is still $V_C I_C$ and this power must all be dissipated in the transistor. Thus to obtain a class A output of say 1 watt it is necessary to use a transistor capable of dissipating 2 watts.

The a.c. collector load resistance is given by

$$R_L = \frac{V_C}{I_C} = \frac{V_C^2}{2P_o}$$

For an output of 1 watt with a supply voltage of 12 volts, the optimum collector load resistance will be 72 ohms, and we must use a transistor with a collector voltage rating of 24 volts and a power rating of 2 watts. If the actual load resistance is 3 ohms, the output transformer turns ratio will be $\sqrt{(72/3)}$, i.e. approximately 5 to 1.

Class A output stages often use an emitter resistance R_E to help in bias stabilisation and to limit peak current. Since power will be dissipated in the resistor, the efficiency of the stage will be reduced, and the maximum output power for a given collector supply voltage will be reduced.

Class A push–pull amplifiers

Distortion caused by non-linearity of the transistor characteristics can be reduced to some extent by the use of the push–pull circuit as shown in Fig. 9.4. The operating conditions of each transistor are the same as those for the single-ended amplifier, and twice the output power will be obtained, i.e. with two 2 watt transistors, a power output of 4 watts will be possible. The even order harmonics are eliminated in the output as long as the two transistors have similar characteristics. However, as with the single-ended class A circuit, the current drain from the supply remains constant even when the signal falls to low levels, and so the efficiency with small signals is very low and at maximum is less than 50 per cent. Because of these factors, the class B push–pull stage is generally preferred and is widely used for high output powers.

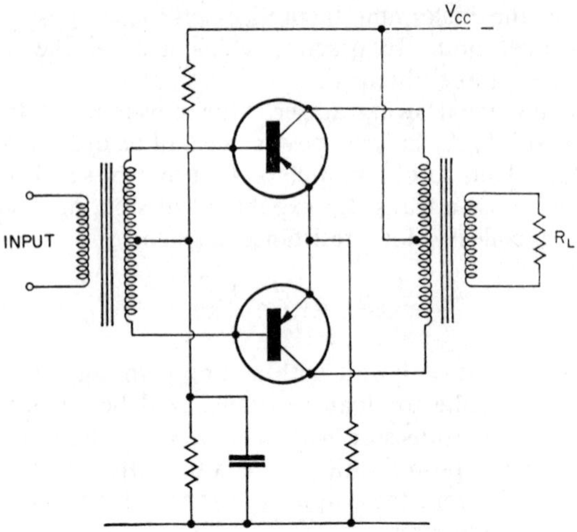

Fig. 9.4.—*Class A push-pull stage.*

Class B push–pull output stage

Class B operation offers several advantages over class A. The efficiency is high under all conditions of input power, and the current drain from the supply is very small at zero signal. As stated earlier, the operating bias point is close to cut-off as indicated in Fig. 9.1 so that under zero signal conditions the voltage across the transistor will be the full supply voltage and the current through the transistor will be almost zero. As the signal is increased, during the positive swing, the current through one transistor increases and the voltage across it falls as the operating condition moves along the load line. During the negative swing, the current increases through the other transistor. The basic circuit arrangement is shown in Fig. 9.5. An important requirement for class B push–pull stages is that the two transistors must have identical current gains, so that the two halves of the waveform are amplified equally.

The maximum power output from the circuit will be the product of the r.m.s. values of the maximum voltage swing and the maximum current swing, i.e.

$$P_o = \frac{V_{CC}}{\sqrt{2}} \times \frac{I_{max}}{\sqrt{2}} = \frac{V_{CC} I_{max}}{2}$$

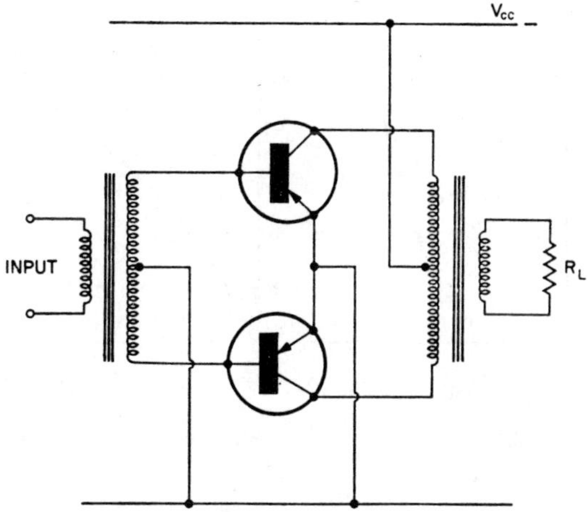

Fig. 9.5.—*Basic Class B push-pull stage.*

The power supplied from the collector supply will be

$$P_S = V_{CC} \times \text{average current}$$

$$= V_{CC} \times I_{max} \frac{2}{\pi}$$

$$= 0{\cdot}636 V_{CC} I_{max}$$

Thus the maximum theoretical efficiency

$$n = \frac{0{\cdot}5 V_{CC} I_{max}}{0{\cdot}636 V_{CC} I_{max}} = 78{\cdot}5 \text{ per cent.}$$

In practice efficiencies of 70–75 per cent can readily be obtained. With an efficiency of 75 per cent a class B amplifier using two transistors each having an allowable dissipation of 2 watts will give an output of 4 × 75/25 = 12 watts.

The load resistance in each collector circuit for a power output P_o will be

$$R_L = \frac{V_{CC}}{I_{max}} = \frac{V_{CC}^2}{2P_o}$$

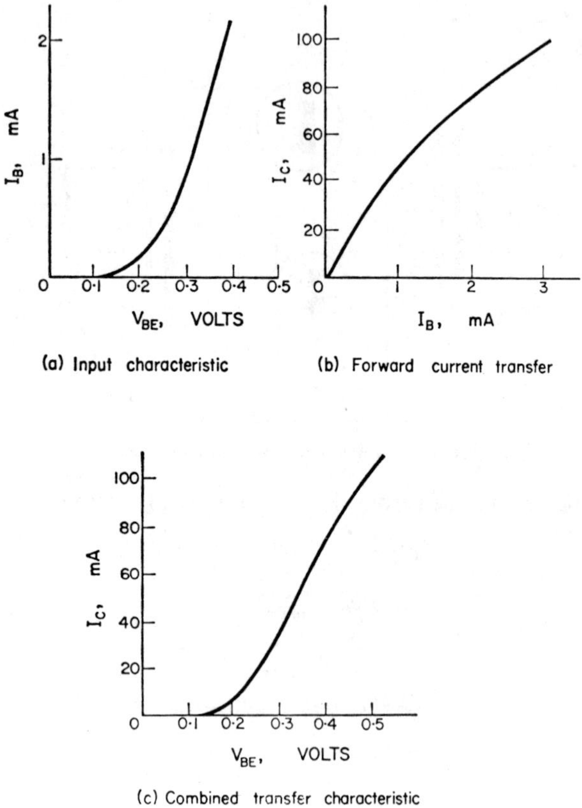

Fig. 9.6.—*Transfer characteristic for a Class B stage.*

With the circuit shown in Fig. 9.5 the collector current with no signal is zero. For a sinusoidal input signal the waveform of the output signal can be deduced from consideration of the combined effect of the input characteristic and current transfer characteristic of the transistors. The input characteristic is shown in Fig. 9.6 (a). This is effectively the forward characteristic of the emitter–base diode, and is very non-linear near the origin. The forward current transfer characteristic is shown in Fig. 9.6 (b); at high currents, the curve flattens due to the current gain falling off. Fig. 9.6 (c) shows the combined base voltage to collector current transfer characteristic.

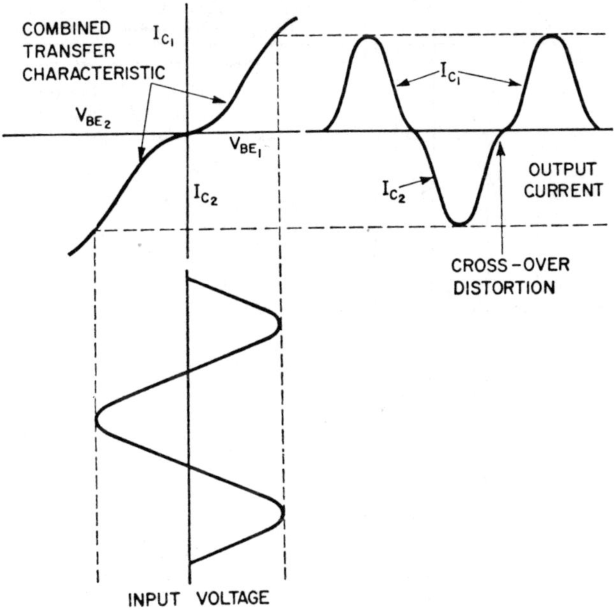

Fig. 9.7.—*Cross-over distortion.*

The shape of the output current waveform due to these effects is shown in Fig. 9.7. It will be seen that serious distortion occurs near the origin. This is known as *cross-over distortion* and as the signal becomes smaller its effect becomes worse. In order to minimise the cross-over distortion, a small forward bias is applied to the transistors so that they pass a small collector current with no signal. The effect of this bias is shown in Fig. 9.8. With germanium transistors, a bias of 0·15 volt is sufficient and with silicon transistors about 0·6 volt. Thus practical class B amplifiers do not operate in the strictly defined class B condition, as the transistors are biased into

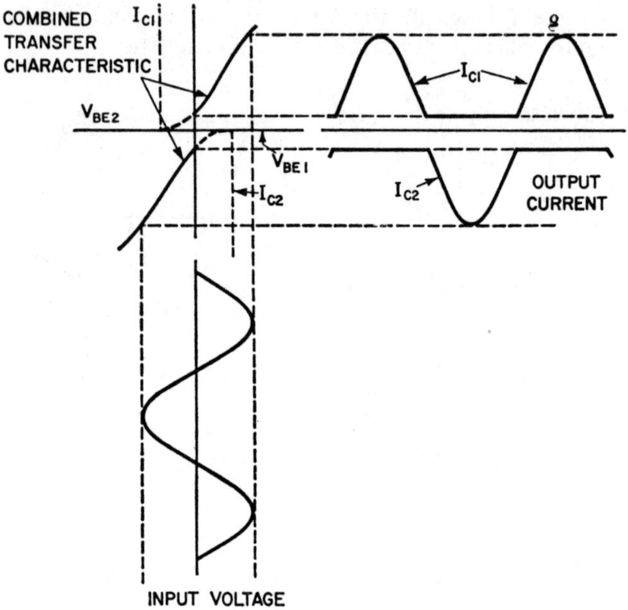

Fig. 9.8.—*Effect of forward bias.*

slight conduction. Fall-off of current gain causes rounding of the peaks of the output waveforms, and the transistors must be chosen so that the distortion produced is limited to an acceptable level. The push–pull connection reduces the even order harmonics, and residual distortion can be reduced by the use of negative feedback.

A complete class B stage is shown in Fig. 9.9. The bias to overcome the cross-over distortion is obtained from a low resistance potentiometer R_1+R_2. The resistor R_2 is in series with the signal and so will cause a slight signal loss. It cannot be by-passed with a capacitor since the input signal rectification by the transistors would cause the capacitor to be charged to give a reverse voltage which would tend to oppose the bias and reintroduce the cross-over distortion. Thus the value of R_2 must be kept as low as possible.

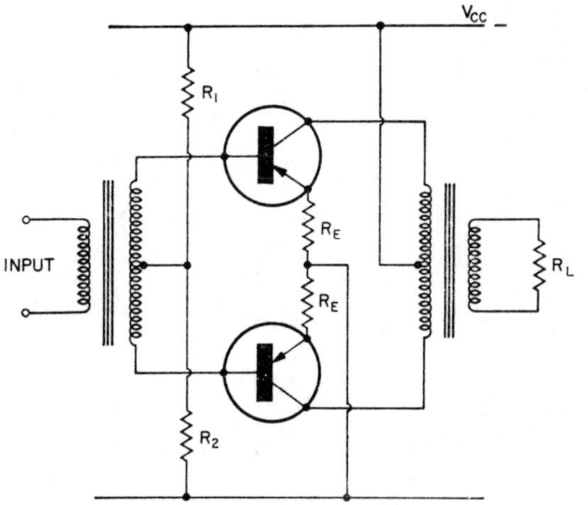

Fig. 9.9.—*Class B stage with bias.*

Small resistors R_E are included in the emitters to help to establish the quiescent operating bias point and to stabilise the transistors against thermal runaway. They also introduce a small amount of negative feedback and so reduce distortion.

Output coupling

It was shown above that with a class B amplifier the load resistance for a power output of P_o is $V_{CC}^2/2P_o$ in the collector circuit of each transistor. It will be seen that this is not dependent on the characteristics of the output transistors—only on the ratings. Fig. 9.10 shows how the output power varies with load resistance for various supply voltages.

Audio power amplifiers are usually required to feed loudspeakers with coil impedances in the region of 3 to 15 ohms. For low power outputs, up to about 2 watts, transformer output coupling is needed to reflect the optimum resistance back from the load to the collector circuit, but for higher

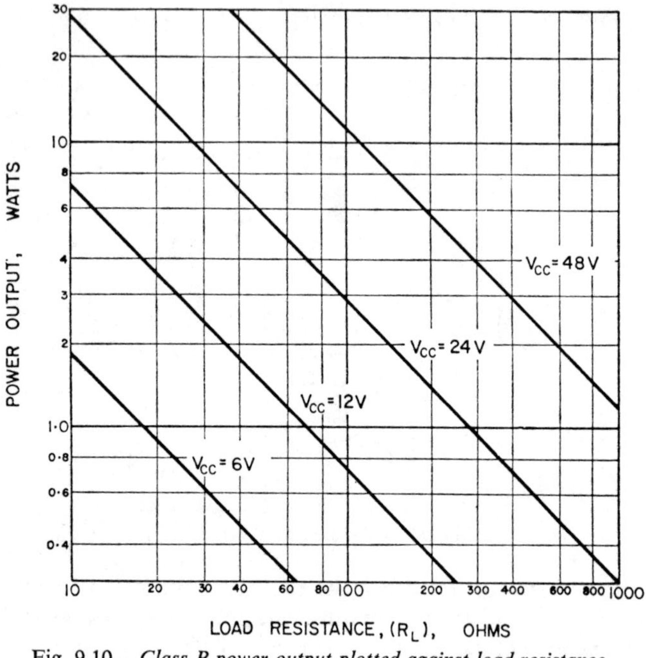

Fig. 9.10.—*Class B power output plotted against load resistance.*

powers the loudspeaker coil resistance is of the same order as the optimum load resistance and so it can be connected direct in the output circuit, as will be shown later.

For the transformer coupled output (Fig. 9.9) the transformer must be designed to have the correct turns ratio. For example, consider a 0·5 watt amplifier with $V_{CC} = 12$ volts feeding a 3 ohms speaker. The amplifier will require a load resistance of 144 ohms in each collector, and so the impedance step down will be 46+46 to 1. The impedance ratio is proportional to the turns ratio squared, and so the turns ratio will be 6·8+6·8 to 1.

The load can be connected directly in the collector circuit as shown in Fig. 9.11. The two output transistors are connected in series to a centre tapped d.c. supply. The inputs are fed from separate windings on the input transformer and are applied to

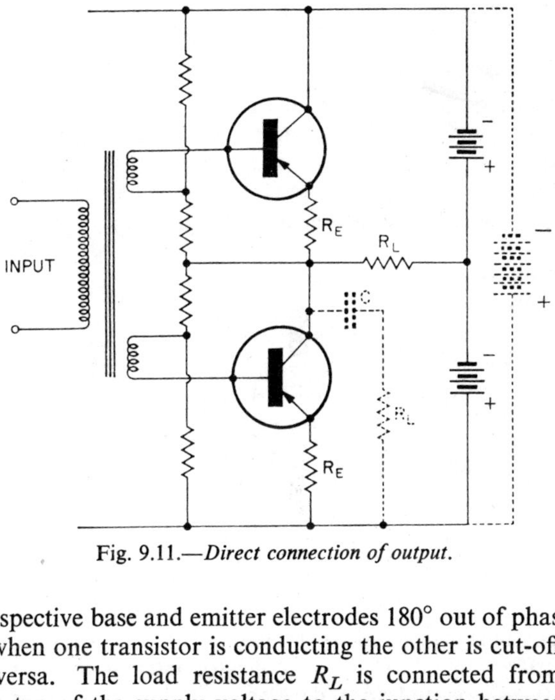

Fig. 9.11.—*Direct connection of output.*

the respective base and emitter electrodes 180° out of phase, so that when one transistor is conducting the other is cut-off and vice versa. The load resistance R_L is connected from the centre tap of the supply voltage to the junction between the two transistors so that it is common to both transistor circuits. Thus the current through R_L will reverse for each half cycle to produce the a.c. output. The transistors are biased by separate bias circuits. It will be seen that the simplicity of the direct load connection is obtained at the expense of the complication of a centre tapped supply voltage. However, it saves the cost of a transformer and avoids the power loss, frequency limitations, weight and space associated with even the best designed transformers.

The centre-tapped supply can be avoided by coupling the a.c. output via an isolating capacitor C as shown dotted in Fig. 9.11. Since the two transistors are normally a matched pair, the d.c. supply voltage will automatically divide equally between them.

Complementary push–pull

Advantage can be taken of the fact that transistors can be either of the two complementary types—pnp and npn. One particularly useful feature is that of obtaining push–pull amplification with single-ended circuits. The basic arrangement, known as *complementary push–pull*, is shown in Fig. 9.12.

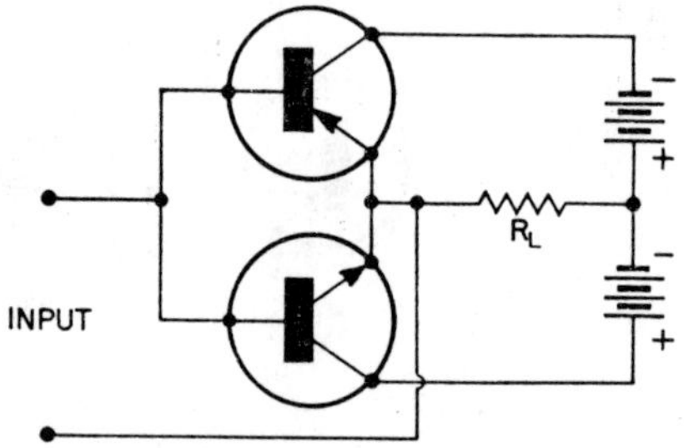

Fig. 9.12.—*Basic complementary symmetry push-pull circuit.*

The input signal is fed to the two base electrodes in parallel. A positive signal results in collector current flowing in the npn transistor and so a positive output voltage is produced across the load resistance R_L. When the input signal is negative the pnp transistor conducts and produces a negative voltage across R_L. Thus if an a.c. signal is applied to the input an amplified signal is developed across the load, but with no signal, no current flows in the transistors, and so we effectively have class B push–pull amplification with a single-ended circuit. As with the normal class B circuit, a small bias must be applied to reduce cross-over distortion. A practical circuit is shown in Fig. 9.13(a). R_2 must be kept as low as possible to

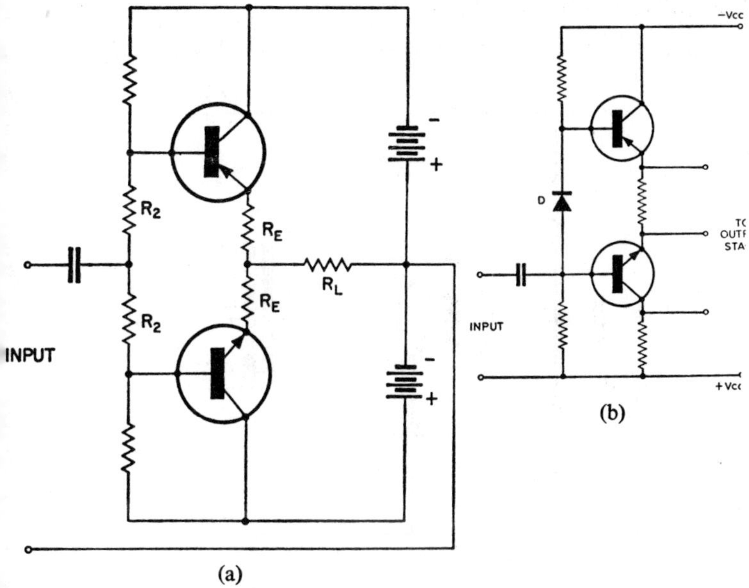

Fig. 9.13.—(a) *Complementary symmetry push-pull stage with bias.* (b) *Complementary pair driver stage with diode bias.*

avoid loss of input signal. The load can be connected directly in the circuit as shown with R_L or can be matched in by means of an output transformer.

The resistors R_2 can be replaced by forward biased diodes to give better bias stabilisation. When germanium transistors are used, a single forward biased silicon diode, *D*, is convenient as shown in Fig. 9.13 (b), which illustrates a typical arrangement for a complementary pair driver stage. The volt-drop across the single silicon diode is of the order required to give suitable bias conditions for the two germanium base-emitter diodes in series. With silicon transistors, two silicon diodes in series will allow the correct order of bias.

Driver stages

In many cases the power input required for an output stage is sufficiently large that the driver stage must also be considered and designed as a power stage. A single-ended class A

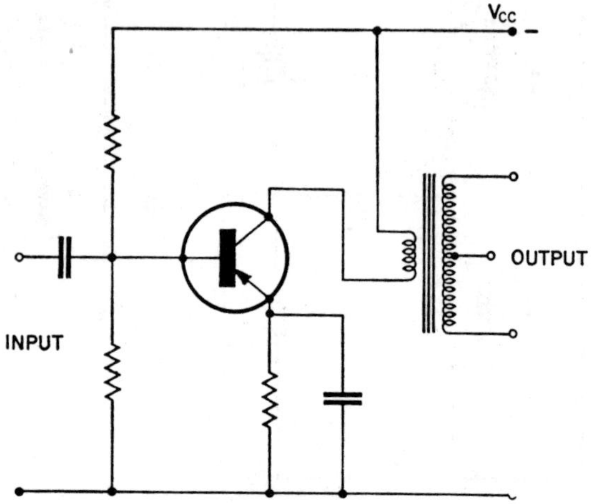

Fig. 9.14.—*Class A driver stage.*

stage is commonly used, with the signal output coupled to the push–pull stage by means of a transformer with a push–pull secondary winding. As with the output transformer, it is necessary to design the coupling transformer so that the input impedance of the output stage is reflected to give the optimum resistance in the collector circuit of the driver transistor. The general arrangement is shown in Fig. 9.14.

Alternatively resistance-capacitance coupling can be used with a split load phase inverting circuit to produce the push–pull driving signal. Two phase inverter circuits are shown in Fig. 9.15. The first gives a low impedance output and the second a high impedance output. By the use of this type of circuit, together with direct connection of the load in the push–pull output circuit, the use of transformers can be completely avoided.

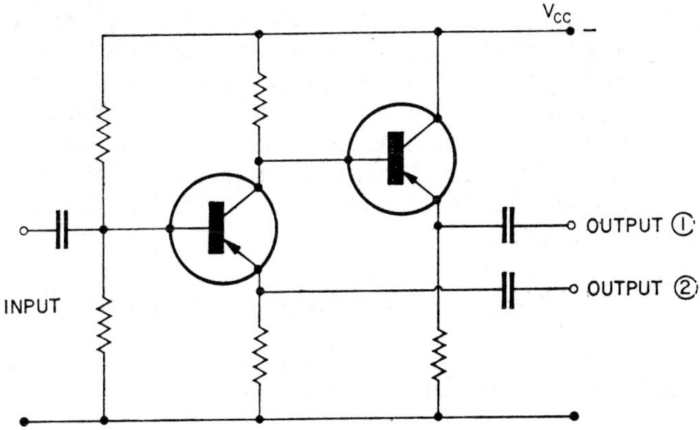

(a) Low impedance output

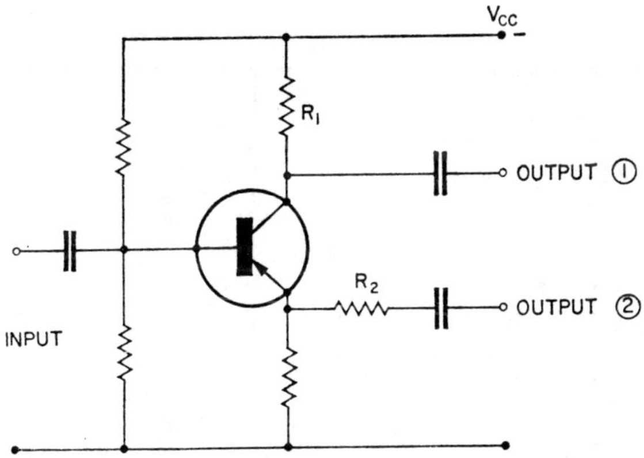

(b) High impedance output

Fig. 9.15.—*Split load phase inverter circuits.* R_2 *in* (*b*) *is used to balance the impedance of the outputs, the collector impedance of the transistor being higher than the emitter impedance. The value of the emitter load resistor can be made higher than the collector load resistor to compensate for signal loss across* R_2.

Negative feedback

With the transformer coupled output, negative feedback is readily applied from the output back to any stage of the amplifier, since the phase of the feedback signal can be chosen at will by arranging the connections to the output winding. It is customary to apply feedback over the output plus driver stages as shown in Fig. 9.17. With a direct coupled load, there is some limitation in that the phase of the output voltage is fixed, and so the feedback must be designed to fit in with the phase requirements. Typical arrangements are included in the practical examples.

TYPICAL AUDIO POWER AMPLIFIERS

To illustrate the various topics discussed in this chapter, several practical examples will be described.

800 mW audio amplifier

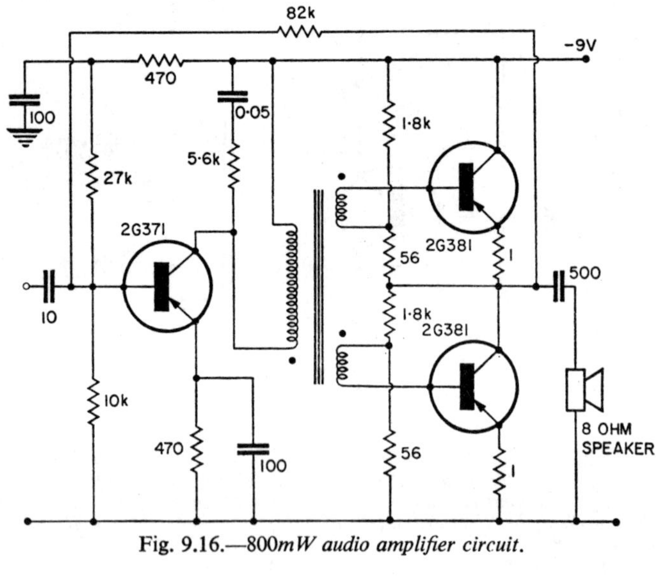

Fig. 9.16.—800*mW audio amplifier circuit.*

This amplifier, shown in Fig. 9.16, is suitable for the audio section of portable radio receivers. The driver stage is a conventional bias stabilised class A stage, coupled to the class B output stage by a transformer with two secondary windings. The output stage uses a matched pair of pnp transistors connected in series across the supply, the 8 ohm loudspeaker being fed direct from the centre point via a coupling capacitor. Feedback is taken from the output to the base of the first transistor, the feedback resistor being selected to give 6 db of negative feedback. The circuit gives 800 mW output with less than 5 per cent distortion. An input of 4 mV will give an output of 50 mW.

1 watt amplifier

Fig. 9.17 shows an amplifier circuit using a complementary push-pull output stage which will give 1 watt output into an 8 ohm loudspeaker using a supply voltage of 9 volts. The circuit

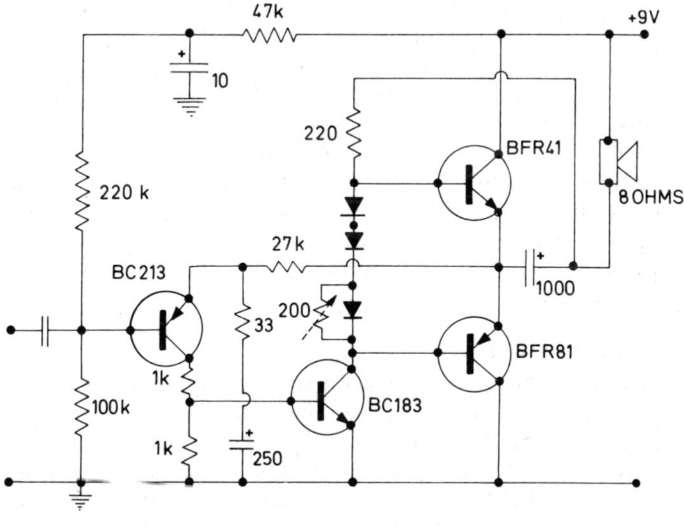

Fig. 9.17.—1 *watt audio amplifier.*

uses silicon planar transistors. The input signal is amplified by a pnp transistor stage and then fed to an npn driver stage which drives the output complementary pair. Biasing of the output transistors to minimise cross-over distortion is accomplished by using three silicon diodes in series with one shunted by a preset resistor which is adjusted to give the biasing voltage necessary for minimum distortion. Negative feedback is applied from the output to the emitter of the input amplifier stage. The amplifier will give one watt output for about 50 mV input.

3 watts amplifier

To obtain higher powers of a few watts, complementary driver transistors can be used with a higher supply voltage as shown in Fig. 9.18. These driver transistors give higher base currents into the output transistors to result in the increased collector currents required for the higher power. The input signal is amplified by an npn stage which feeds a first driver transistor (pnp) the output of which in turn feeds the complementary driver transistors for the output stage. The first driver stage collector load is connected to the negative line via the loudspeaker; this arrangement allows a higher value of load resistor and so a lower current through the first driver transistor while still maintaining the required voltage swing. Biasing for minimum cross-over distortion is similar to that used in the 1 watt amplifier and negative feedback is again from the output back to the emitter of the input stage. An output of 3 watts is obtained with about 80 mV input. Harmonic distortion is less than 1%.

20 watts amplifier

A still higher power amplifier, capable of delivering 20 watts into 8 ohms is shown in Fig. 9.19. The output and driver stages are similar to those of the 3 watt amplifier except that higher power transistors are used with a higher voltage supply—40 volts. An alternative method of biasing for minimum cross-over distortion is used, the voltage across a transistor being set for minimum distortion by adjusting its base potential. The input

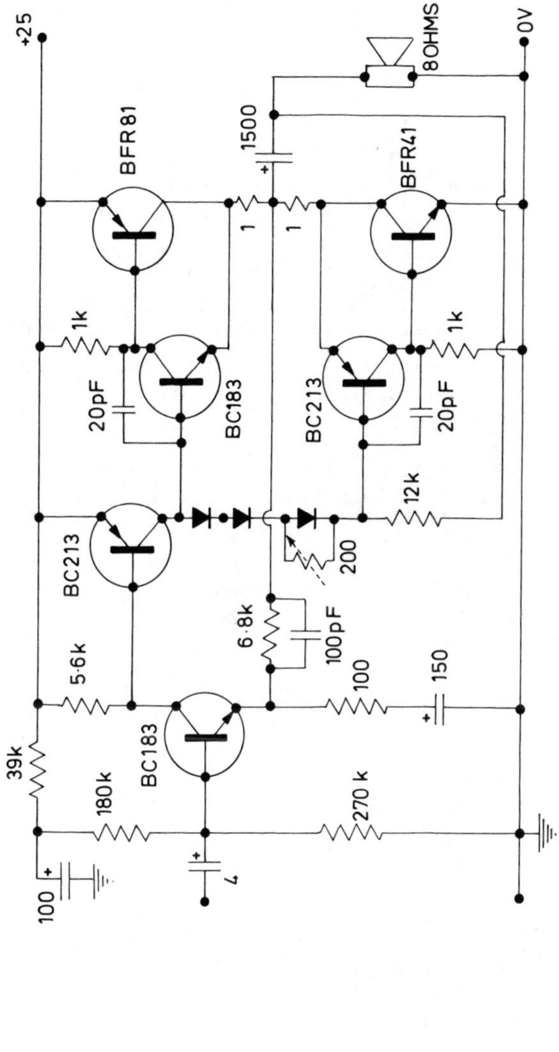

Fig. 9.18.—3 watts audio amplifier.

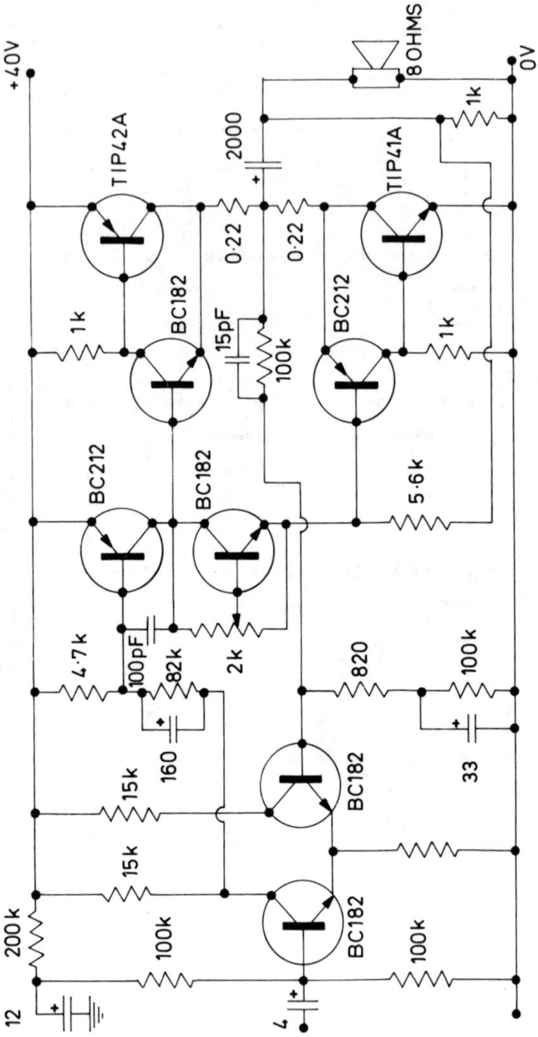

Fig. 9.19.—*20 watts audio amplifier.*

amplifier stage uses two npn transistors in a long tailed pair arrangement. The signal is amplified by the first transistor and negative feedback is fed in through the second. 20 watts output is obtained with about 100 mV input and harmonic distortion for full output is less than 0·2%.

10
HIGH FREQUENCY AMPLIFIERS

THE high frequency performance of junction transistors is limited by several factors that are negligible at low frequency. The most important of these are: (1) The transit time of current carriers in passing from emitter to collector. (2) The dispersion of this transit time. (3) Emitter-base junction capacitance. (4) Collector-base junction capacitance. (5) Collector capacitance-base resistance feedback.

Transit time

After being injected from the emitter into the base region, the minority carriers (holes in pnp transistors, electrons in npn transistors) flow to the collector by a process of diffusion. Such diffusion takes places due to a difference in the density of the carriers, the flow being from a region of high density near the emitter junction to a region of low density adjacent to the collector junction, or more accurately adjacent to the edge of the collector depletion layer in the base region. In the alloy transistor, the base region between the emitter and the collector layer is all at the same potential, and so there is no electric field to influence the carriers. Thus the carriers flow by diffusion until they reach the collector depletion layer, where they encounter an electric field which attracts them across the collector junction into the collector region. The time taken for the carriers to flow across the base region is $t = w^2/D$ seconds, where w is the width of the base region in cm and D is the diffusion constant for the carrier in cm²/sec.

For a typical low frequency germanium alloy transistor, $w = 0{\cdot}0005$ inch (0·0012 cm) and $D = 33$ cm²/sec for holes in n-type germanium. Thus

$$t = \frac{(0{\cdot}0012)^2}{33} = 0{\cdot}04\ \mu\text{sec}$$

This gives, ignoring for the moment all other factors, an upper frequency limit of 12·5 Mc/s.

Since this maximum frequency depends upon the square of the thickness of the base region, it is readily apparent that the first requirement for a high frequency transistor is a very thin base region. A germanium mesa transistor for a maximum frequency of say 1,250 Mc/s will have a transit time of 0·0004 μsec and requires a base width of 0·00005 inch.

Dispersion of carrier transit time

The flow of carriers through the base by diffusion is essentially a random process, an individual carrier experiencing many collisions and changing direction of motion many times

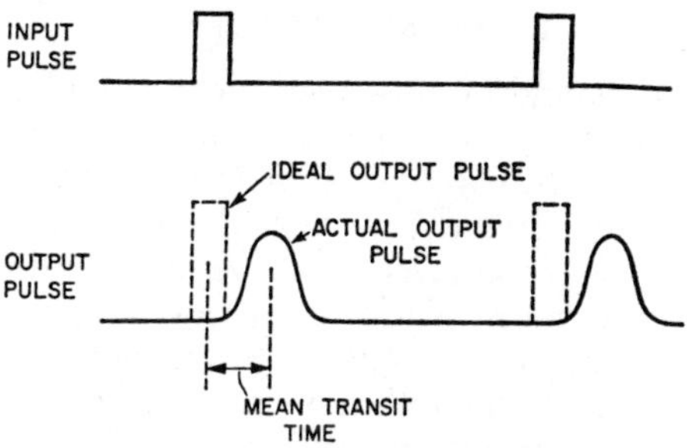

Fig. 10.1.—*Transit time and dispersion.*

during its travel from the emitter to collector. The value for the diffusion constant is an average value, and the transit time for individual carriers will vary considerably, giving rise to a spreading, or dispersion, of the output signal at the collector.

This effect can be illustrated by considering what happens to a very short input current pulse. The current pulse in the collector will not only be delayed by the average transit time, it will also be spread out in time by the dispersion, resulting in a reduction in amplitude and so a reduction in the effective current gain. This is illustrated in Fig. 10.1.

Such dispersion can also be introduced if the spacing between the emitter and collector is not constant over the whole area, giving different path lengths through the base region. Thus it is very important that the emitter and collector junctions are both flat and parallel.

Emitter to base capacitance

During the diffusion of carriers across the base region due to a given signal input, the number of carriers present in the base region at any moment will depend upon the concentration gradient of the carriers. If the signal is increased, a certain number of carriers must flow into the base region to establish a higher concentration before greater current can start to flow. This is similar to charging a capacitance, and the effective capacitance due to this effect is called the *emitter diffusion capacitance.* Its value can be approximated to

$$C_{be} = 2{\cdot}8 \times 10^3 w^2 I_e \text{ pF}$$

when w is in thousandths of an inch and I_e in milliamperes.

For the typical germanium alloy transistor, with $w = 0{\cdot}5$ thousandths of an inch and $I_e = 10$ mA, $C_{be} = 7{,}000$ pF.

There is also an emitter to base depletion capacitance, $C_{e(\mathrm{dep})}$, formed by the depletion layer effect. At zero emitter current the diffusion capacitance will be zero, and the emitter capacitance will consist entirely of the depletion component, but at high emitter currents the diffusion capacitance predominates and the depletion capacitance can be neglected.

Collector capacitance

The collector–base capacitance is formed by the junction depletion layer as described in Chapter 2. It is directly proportional to the area of the junction and is inversely proportional to the square root of the voltage across the junction for an abrupt junction, as in the alloy transistor, or to the cube root of the voltage for a graded junction as in the diffused transistor.

High frequency equivalent circuit

With the above factors in mind, it is now possible to see how the low frequency equivalent circuit in Fig. 6.3 is modified at high frequency. First consider the emitter current flowing through the emitter junction and on through the base region (Fig. 10.2). The depletion capacitance $C_{e(\text{dep})}$ will appear across r_e, the forward resistance of the emitter junction. The

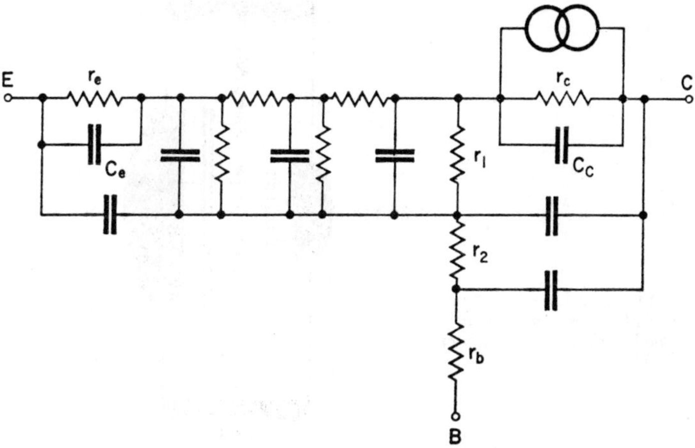

Fig. 10.2.—*High frequency equivalent circuit.*

diffusion capacitance can be represented by a resistance–capacitance network as shown, since the concentration gradient must be built up in a distributed manner through the base region. On the collector side, the collector depletion layer $C_{c(\text{dep})}$ appears across the resistance of the reverse biased collector junction. In considering what happens to the base current, note must be taken of the distributed nature of the emitter and collector capacitances across the junction areas. Fig. 10.3 shows a diagrammatic representation of the emitter and base capacitances distributed along the ohmic base resistance. There are three components of the base resistance, r_b outside the collector junction, r_2 under the collector junction only and r_1 under both junctions. These various components of resistance and capacity can be approximated as shown in Fig. 10.2.

As in the low frequency equivalent circuit, a current generator must be included in the collector circuit. This will still be proportional to the emitter current i_e, but account must be taken of the fact that the current transfer ratio falls off with increasing frequency. If the frequency at which the current

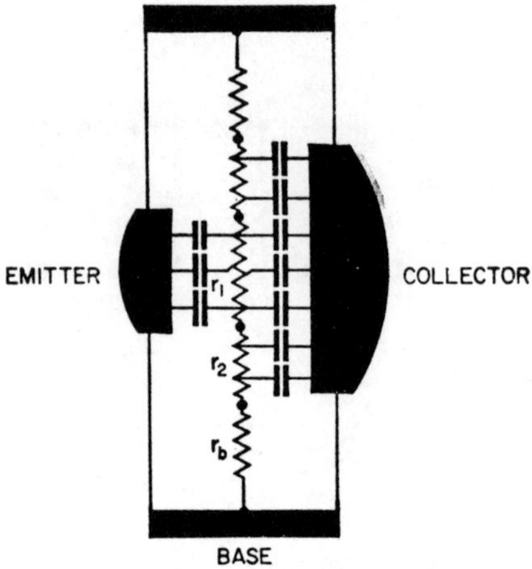

Fig. 10.3. — *Schematic representation of base resistance and junction capacitance.*

transfer ratio has fallen by 3 db is designated f_α (the α cut-off frequency) then an approximate expression for the transfer ratio at any other frequency f is

$$\alpha = \frac{\alpha_o}{1 + jf/f_\alpha}$$

where α_o is the current transfer ratio at low frequency $[j = \sqrt{(-1)}]$. This expression also takes account of the fact that there is an effective phase change between the emitter current and collector current due to the transit time. Thus the current generator in the high frequency equivalent circuit has the value

$$\frac{\alpha i_e}{1 + jf/f_\alpha}$$

For the purpose of a general understanding of the high frequency operation, it is possible to approximate the equivalent circuit as shown in Fig. 10.4. Since the effective emitter junction capacitance appears across a very low resistance (the forward resistance of a germanium emitter junction is approximately equal to $25/I_e$, where I_e is in milliamperes) it can be ignored at medium frequencies, but as the frequency is

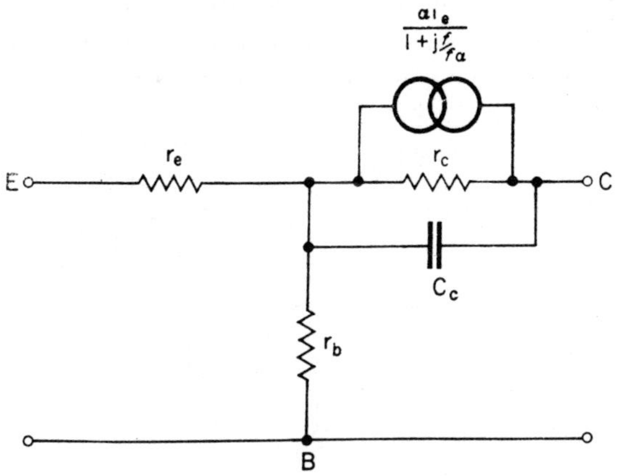

Fig. 10.4.—*Simplified high frequency common base equivalent circuit.*

increased it becomes more important. Of major importance is the combined effect of collector capacitance and base resistance. An output signal appearing across the collector load causes a current to flow back through C_c into r_b and the voltage produced across r_b appears in the input circuit. The phase relationship between the output signal and the feedback signal depends upon the nature of the load, and on the values of C_c and r_b. Also, as stated above, at high frequencies there is a phase shift of the forward current transfer, and under certain conditions the combined effect is to produce positive feedback which results in instability or unwanted oscillations.

The magnitude of the feedback voltage appearing across r_b is given by

$$v_r = V_o \frac{r_b}{(r_b + 1/jwC_c)} = V_o \cdot \frac{jwr_b C_c}{(1 + jwr_b C_c)}$$

and to keep this low it is necessary to keep the product $r_b C_c$ as low as possible. In practice, the $r_b C_c$ product usually limits the high frequency performance before the transit time effect discussed earlier.

Common emitter high frequency equivalent circuit

The high frequency equivalent circuit can be presented for the common emitter connection as shown in Fig. 10.5. Again,

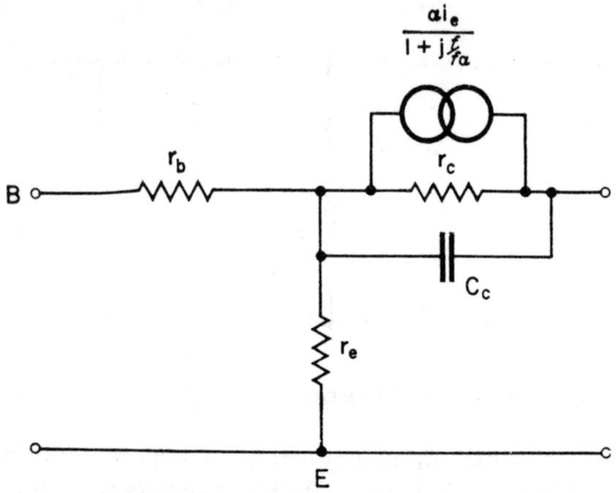

Fig. 10.5.—*Simplified high frequency common emitter equivalent circuit.*

at high frequencies the collector capacitance causes signal feedback. The variation of common emitter current gain with frequency is expressed as

$$\beta = \frac{\beta_o}{1+jf/f_\beta}$$

where β_o is the low frequency gain and f_β the frequency at which it has fallen by 3 db. The relation is plotted in the graph shown in Fig. 10.6 (a). The slope of the fall-off is 6 db per octave. Also shown, in Fig. 10.6 (b), is the relation for α.

Instead of quoting the cut-off frequency as f_α or f_β, it has now become customary to quote an alternative parameter f_T. This

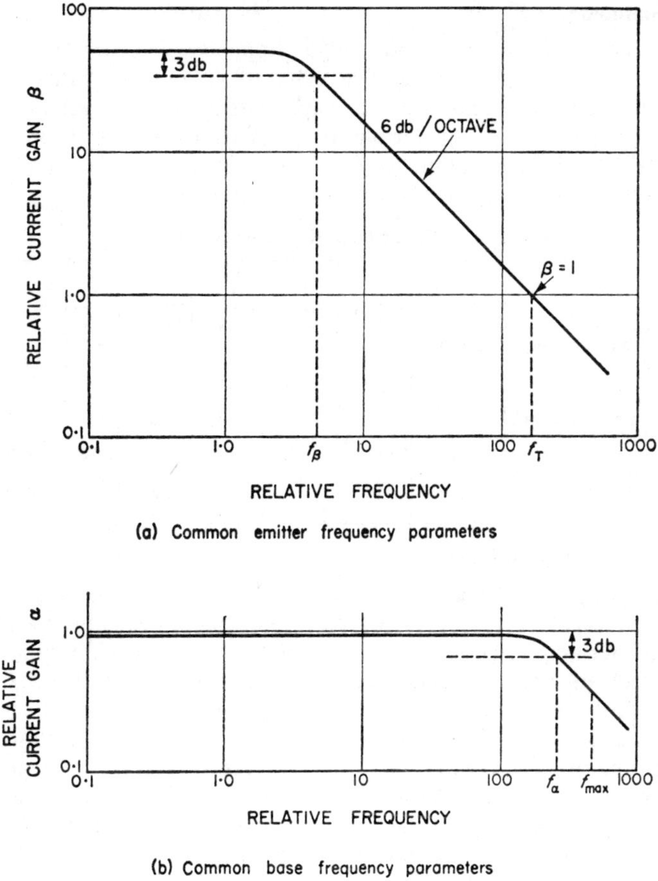

Fig. 10.6.—*Frequency parameters.*

is the frequency at which the common emitter current gain has fallen to unity and is usually determined by measuring the current gain at a frequency above f_β and extrapolating at 6 db per octave. For example, a transistor with a current gain of 3 at 20 Mc/s will have an f_T of 60 Mc/s.

Summary

Summarising, for good high frequency performance a transistor should have a high value of f_T and low values of base resistance and collector capacitance. Typical values of these parameters for various types of transistor are given in Table 10.1. The parameter f_{max} is the maximum frequency of oscillation, and is mentioned in Chapter 11.

Table 10.1. Transistor high frequency parameters

Class of transistor	*Type*	f_β (Mc/s)	f_T (Mc/s)	f_α (Mc/s)	r_b (ohms)	C_c (pF)	f_{max} (Mc/s)
Germanium alloy	2G304	0·12	12	18	90	8·5	25
Germanium alloy-diffused	2G414	2	120	180	16	2·5	345
Silicon epitaxial planar	2N743	8	300	450	40	5	750
Silicon epitaxial planar	TIX3016	50	1,700	2,500	3·5	1·4	3,700
Germanium diffused mesa	2N2999	50	1,400	2,000	2·5	1·5	3,850

Tuned h.f. amplifiers

If a suitable transistor is chosen, high frequency signals fed to the base of the transistor will be amplified and will appear with greater amplitude in the collector circuit. It is a usual requirement of high frequency circuits that only a certain well-defined band of frequencies is amplified. Selection of the required band of frequencies is arranged by means of a tuned circuit in series with the collector. As with low frequency amplifiers the output impedance of the transistor is relatively high and input impedance of the next stage is low. For efficient coupling the interstage circuit must be arranged for maximum power transfer, that is, the output impedance of the first stage must be matched to the input impedance of the next stage. This can conveniently be accomplished by means of step-down inductive coupling from the tuned circuit inductance.

Single tuned circuit

The simplest tuned amplifier arrangement is shown in Fig. 10.7. The collector feeds a single tuned circuit, with a secondary winding tightly coupled to the tuned circuit inductance having a turns step-down ratio to give the correct impedance matching. A resistance R across the tuned circuit will match to R/n^2 across the secondary winding, where n is

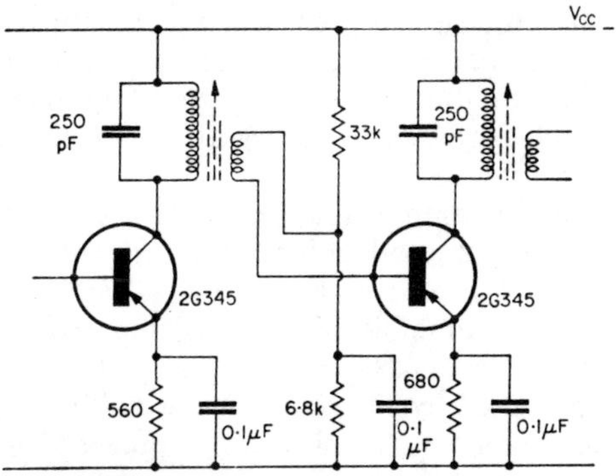

Fig. 10.7.—*Single-tuned amplifier stage.*

equal to the turns step-down ratio. Under matched conditions the input impedance of the second transistor will be reflected back as a resistance equal to the output resistance of the first transistor, and so the effective resistance across the tuned circuit is $0{\cdot}5\,R_o$. The bandwidth (B) of a tuned circuit to the 3 db down points is equal to the mid frequency (f_o) of the resonance divided by the loaded Q of the tuned circuit, i.e. $B = f_o/Q_L$. Assuming that the inductance has zero resistance, the loaded Q of the tuned circuit will be given by the total parallel resistance divided by the reactance of the tuning capacitance at resonance, i.e.

$$Q_L = \frac{0{\cdot}5R_o}{1/wC} = 0{\cdot}5wCR_o$$

In practice, the Q will be smaller than the value given by this expression because of the series resistance of the tuning inductance. If the bandwidth in a practical circuit is too low, an additional parallel damping resistance can be connected across the tuned circuit to reduce Q_L. If the bandwidth is too great, it can be reduced by making C larger, and reducing the inductance correspondingly to restore resonance. The increase of C will reduce the bandwidth. The secondary winding must also be reduced to maintain the correct turns ratio. Alternatively the collector can be tapped down the tuned circuit, so that the effective parallel resistance across the tuned circuit is increased, thus resulting in a higher Q_L. A typical arrangement is with the collector tapped across about 75 per cent of the winding.

For a practical amplifier stage, with a transistor output resistance of 22k ohms and an input resistance of 600 ohms, the primary to secondary turns ratio will be $\sqrt{(22{,}000/600)} = 6$ to 1. In a 465 kc/s i.f. amplifier for a medium wave receiver a tuning capacitance of 250 pF is typical. The d.c. bias condition is established by the base voltage stabilisation method described in Chapter 7, the bias being applied to the base via the secondary winding of the coupling transformer. Both the emitter and base voltage resistors must be by-passed with capacitors.

It can be shown that the maximum available power gain of a transistor high frequency amplifier at a frequency f is

$$A = \frac{\alpha_0 f_T}{8\pi r_b C_c f^2}$$

This stresses the importance of a high f_T and a low product of $r_b C_c$. A typical 465 kc/s i.f. amplifier stage with a bandwidth of $\pm 4{\cdot}5$ kc/s will have a gain of about 35 db.

Double tuned circuit

For better selectivity, a double tuned bandpass circuit can be used, as illustrated in Fig. 10.8. The collector can be tapped down the primary coil as in the case of the single tuned circuit, and the matching to the input of the next stage is accomplished by tapping the base connection even further down the secondary coil. If the two coils are similar and the collector is tapped across 75 per cent of the primary winding, the base must be tapped across about 12·5 per cent of the secondary winding. The spacing between the primary and secondary coils must be arranged for the critical coupling to give the required bandpass response.

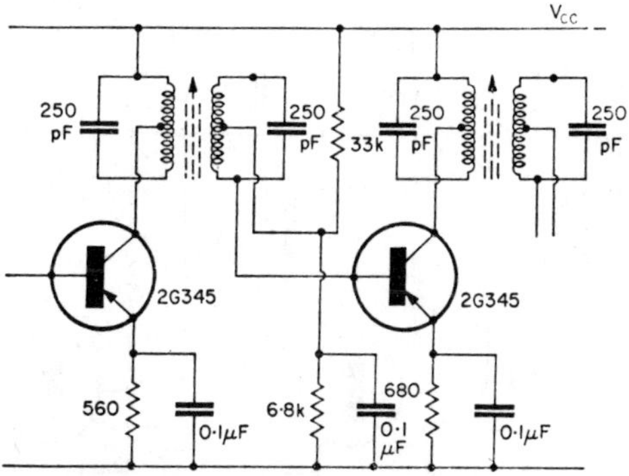

Fig. 10.8.—*Double-tuned amplifier stage.*

Unilateralisation

In tuned amplifiers the transistor internal feedback described earlier causes considerable difficulty in obtaining stability, and even if spurious oscillation does not occur the feedback results in the resonant frequencies of the input and output tuned circuits being interdependent, that is, an adjustment to the output circuit will affect the input circuit and vice versa. This makes it extremely difficult in a multistage amplifier to line up the resonant circuits to the required frequency.

It is possible to neutralise the feedback in a tuned amplifier by means of a process called *unilateralisation.* An external signal, exactly equal in magnitude and opposite in phase, is fed back from the output to the input so that it cancels out the unwanted signal feedback produced internally in the transistor. The amplifier is then said to be unilateralised, that is it will transmit energy in one direction only. It is usual to apply such neutralisation to each individual stage, and then the design and manipulation of a multistage amplifier becomes relatively straightforward.

The external feedback signal can conveniently be obtained from the secondary of the interstage coupling circuit, with the sense of the secondary winding arranged to give the required phase reversal. Then it is only necessary to feed back to the

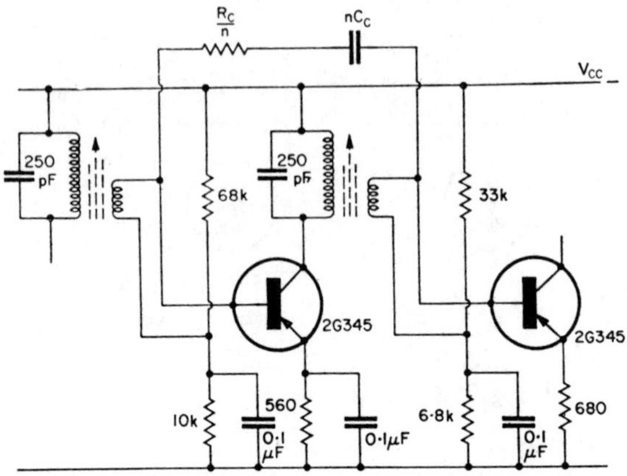

Fig. 10.9.—*Feedback unilateralisation.*

input circuit via a capacitance and resistance to simulate the conditions in the transistor. This arrangement is shown in Fig. 10.9. Since the signal voltage will have been stepped down from the collector winding to the secondary base winding, the impedance of the external feedback circuit must be correspondingly lower than the transistor internal feedback impedance in order to produce the same current in the input circuit. Thus if the turns step-down ratio is n, the external feedback must be nC_c. A resistance may be necessary in series with the capacitance to compensate for the resistive component of the internal feedback.

Automatic gain control

Perhaps one of the most common uses of the tuned amplifier is in the i.f. amplifiers of radio and television receivers. In this application it is required to vary the gain of the amplifier inversely with the strength of the signal received at the antenna, so that the output of the amplifier remains constant. In a receiver without a.g.c., one component of the output of the detector at the end of the i.f. amplifier is a d.c. voltage that is

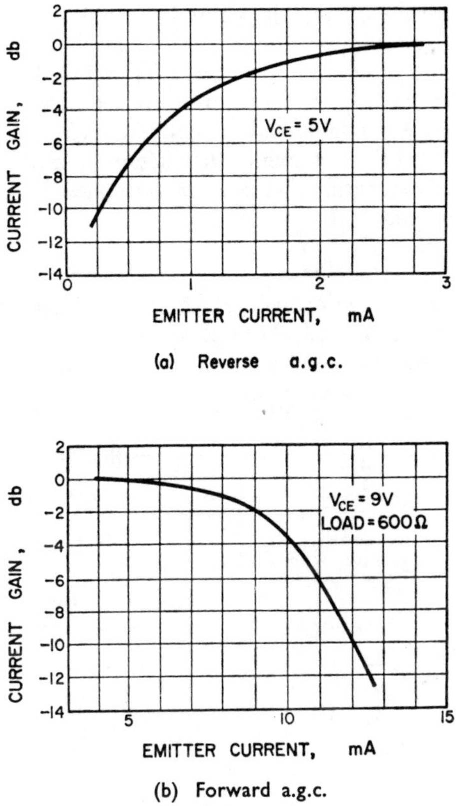

Fig. 10.10.—*A.G.C. characteristics.*

directly proportional to the received signal. This d.c. voltage can be used to vary the gain of the tuned amplifier stages in two ways. The first is called *reverse a.g.c.*, in which the d.c. voltage is fed back to one of the amplifier transistors so as to bias it towards cut-off. As the emitter current of a transistor is reduced the input resistance rises and the gain falls. Thus as the output of the complete amplifier tends to increase, the gain of the controlled stage tends to fall. The second method is called *forward a.g.c.*; in this method the polarity of the d.c. voltage feedback is reversed so that the emitter current increases with the output signal. A resistance is included in the d.c. supply to the collector so that as the emitter current increases the collector voltage falls, and the combined effect results in a decrease in amplification due to the current gain of the transistor falling. The characteristics of the two types of control are illustrated in Fig. 10.10.

Typical 465 kc/s amplifier

A typical two stage 465 kc/s i.f. amplifier illustrating some of the features described above is shown in Fig. 10.11. The arrangement shown is a compromise design suitable for the

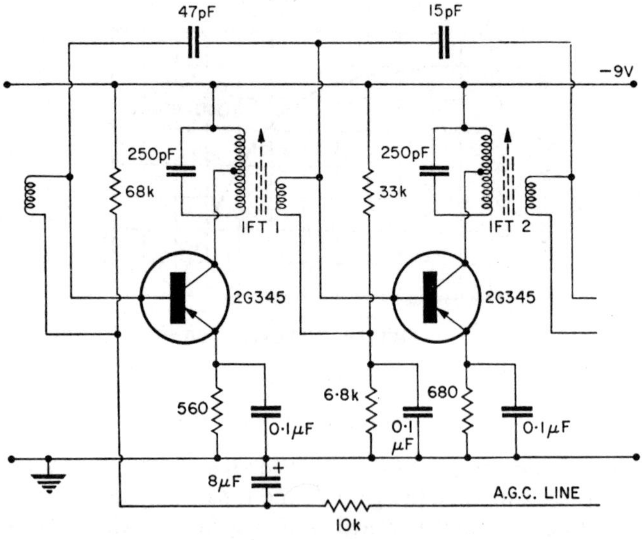

Fig. 10.11.—*Typical 465 kc/s amplifier.*

small portable type of receiver. Single tuned transformers are used, $I.F.T_1$ has a total of 170 turns with the collector tap at 54 turns and a secondary base winding of 9 turns. $I.F.T_2$ has 145 turns with a collector tap at 32 turns and a secondary winding of 17 turns. The tuning capacity for both stages is 250 pF. The first stage is a.g.c. controlled, the second stage has fixed bias, and both stages are neutralised with simple capacitance feedback.

Two-transistor cascode circuit

For higher frequency amplifiers, in the 10–100 Mc/s range, the internal feedback becomes more serious, and an interesting circuit aimed at overcoming the feedback problem is the cascode circuit using two transistors as shown in Fig. 10.12.

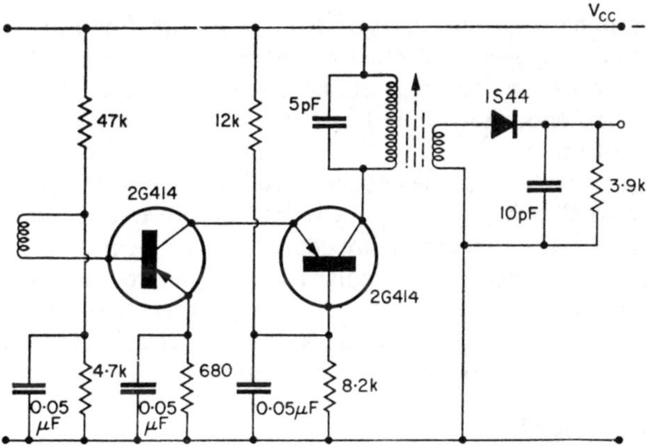

Fig. 10.12.—*Cascode connected* 36 *Mc/s I.F. amplifier stage.*

It consists of a common emitter transistor driving a common base stage. This transistor combination can be considered as a single circuit. The input impedance is that of a common emitter stage and the output impedance the high value of a common base stage. The feedback is the product of the individual values for the two transistors and so is several orders of magnitude lower than for one transistor. In fact, the feedback is so low as to be negligible, and no neutralisation is necessary. Tuned amplifiers using this circuit are quite stable and exhibit a high degree of isolation between output and input circuits. The circuit shown in Fig. 10.12 is one stage of a television receiver i.f. amplifier at 36 Mc/s. It gives a stage gain of the order of 30 db, somewhat higher than a single transistor circuit, this extra gain being obtained at the expense of the extra transistor. The important point, however, is that two inexpensive transistors can be used instead of one more costly type. Also, since two of these cascode stages will give a similar gain to three single transistor stages, one coupling transformer and several passive components will be saved, further offsetting the cost of the additional transistor.

Wide-band tuned amplifiers

Amplifiers in the 10–100 Mc/s range are often used for television receivers and radar receivers, and in both cases wide bandwidths are required, of the order of 5 Mc/s. To obtain this wide bandwidth the Q of the tuned circuit must be very low—of the order of 5 to 10. If a single tuned circuit is used, the turns ratio to the secondary coupling coil can be made near unity so that heavy damping is introduced across the tuned circuit by the low input impedance of the next stage. If a double tuned bandpass circuit is used it may be necessary to connect a suitable resistance across the collector tuned circuit and also to arrange that the secondary circuit is correctly damped by the input of the next stage.

V.H.F. amplifiers

Although unilateralisation generally offers advantage, it has certain limitations, and the present trend in high frequency transistor design is towards lower and lower collector capacitance, so that the need for unilateralisation can be avoided completely up to as high a frequency as possible. However, it appears that some measures are still necessary at v.h.f. and at u.h.f. to avoid trouble due to the feedback. At these high frequencies, unilateralisation becomes more difficult—it may be necessary to adjust the external feedback for each individual transistor and, even then, the neutralisation may only hold over a narrow frequency band.

An alternative to unilateralisation is mismatching, in which the feedback is reduced by overcoupling. This results in a certain loss of gain, but results in a much greater reduction of feedback, and using this method useful and stable gains can be obtained up to v.h.f. and u.h.f. over broad frequency bands. A typical amount of overcoupling is such that the input resistance of a stage is reflected back to the collector of the previous stage as one-half of the transistor output impedance. This will result in a signal power loss of approximately 12 per cent but the voltage feedback will be reduced by the order of 60 per cent.

U.H.F. amplifiers

A typical application of transistors in the u.h.f. band is for the signal tuner/converter for television receivers covering the band of 470–800 Mc/s. One arrangement is shown in Fig. 10.13.

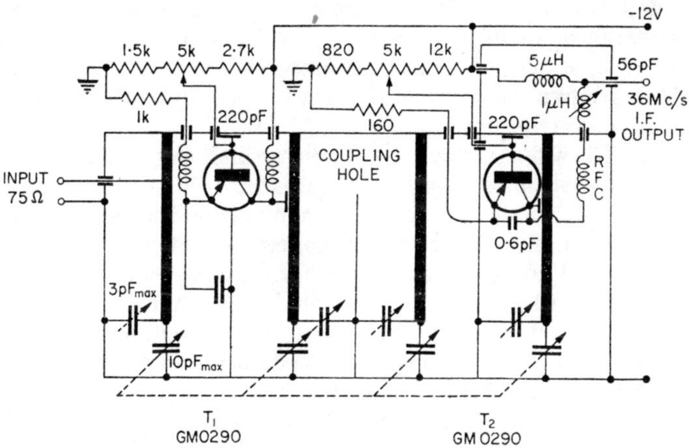

Fig. 10.13.—*Quarter-wave line U.H.F. television tuner circuit.*

Each tuned circuit consists of a quarter-wave rod forming the inner conductor of a coaxial cavity, the rectangular chassis compartment forming the outer body. The circuit uses a common base r.f. amplifier stage (T_1) feeding a common base self-oscillating mixer stage (T_2) via a bandpass circuit. Germanium epitaxial mesa transistors type GM0290 are used. The interstage coupling is a bandpass arrangement with the mutual coupling through a hole in the chassis screen separating the two circuits. The self-oscillating mixer uses a 0·6 pF capacitor from collector to emitter to provide adequate feedback for oscillation. Output at i.f. (36 Mc/s) is developed across a tuned circuit in the collector of the mixer transistor. The r.f. stage gives a stage gain of the order of 18 db with a noise figure of 5–10 db, depending upon the frequency.

Video or pulse amplifiers

In order to amplify television or radar signals a wide-band amplifier is necessary. Such amplifiers are called video or pulse amplifiers, and are required to amplify a band of frequencies extending from a few cycles per second up to the order of 5 Mc/s. Since it is not possible to obtain this frequency coverage with a transformer, the common emitter arrangement is the only possible circuit to give voltage gain. If the basic *RC* coupled common emitter amplifier stage is considered, there are two main effects that cause the gain to fall off with increasing frequency.

(a) The reduction in current gain at high frequencies.

(b) Capacitance shunting of the collector load resistance.

The fall off in current gain is purely a function of the transistor used. The common emitter current gain will have fallen by 3 db at a frequency approximately equal to f_T/β. Thus if a 5 Mc/s response is required with a transistor current gain of 25, a transistor with an f_T of 125 Mc/s will be necessary.

The capacitive shunting of the collector load resistance can be reduced by means of inductive compensation. An inductance is added in series with the load resistance R_L. The value of the inductance is chosen so that as the frequency increases the decrease in capacitive reactance in shunt with the load resistance is compensated by the increase of the reactance of the inductance. Without any inductance, the impedance of the parallel resistance capacity combination will have fallen by 3 db at the frequency at which the capacitive reactance equals the load resistance. If an inductance equal to $\frac{1}{2}CR_L^2$ is connected in series with the load resistance, the fall at that frequency will be reduced to zero.

To take a practical example, consider a video output stage required to develop 100 volts output to feed a cathode-ray tube having a total capacitive load of 25 pF with a frequency response of 5 Mc/s. The load resistance

$$R_L = \frac{1}{2\pi fC} = 1{,}500 \text{ ohms}$$

The compensating inductance $L = \frac{1}{2}CR_L^2 = 22\,\mu\text{H}$. The transistor current swing to give 100 volts output

$$= \frac{100}{1{,}500} = 66 \text{ mA}$$

11

TRANSISTOR OSCILLATORS

ANY amplifier can be converted into an oscillator by feeding back part of the output signal to the input. The conditions for oscillation are: (1) The overall power gain must be greater than unity. (2) The feedback signal must be in phase with the input.

The nature of the oscillation will be determined by the load and feedback conditions. If the load is a tuned circuit, the oscillator will produce a sinusoidal waveform at the resonant frequency of the tuned circuit. If the load and feedback circuits satisfy the required conditions for oscillation over a range of frequencies, a pulse or square type waveform will result. These two categories of oscillation form a convenient division for the subsequent discussion.

SINEWAVE OSCILLATORS

Tuned collector feedback oscillator

The basic tuned collector oscillator circuit is shown in Fig. 11.1. This circuit can be used for a wide range of frequencies, from low audio frequencies using laminated iron cored transformers up to tens or even hundreds of megacycles using r.f. techniques. The values in Fig. 11.1 are suitable for 1 to 2 Mc/s. The feedback signal is obtained by means of a small winding inductively coupled to the collector tuned circuit inductance. Since there is a phase reversal in the transistor when operating in the common emitter mode, it is necessary to ensure that the voltage feedback to the base is phase reversed with respect to the collector by arranging the sense of the feedback winding. The number of turns on the feedback winding is decided such that the overall loop gain from input, through the transistor and back through the feedback circuit to the input, is somewhat greater than unity. The gain can then be finally set by means of a resistor in the emitter circuit. By suitable adjustment of this resistor and its by-passing capacitor, the transistor can be prevented from running into saturation or cut-off, and so ensure good output waveform. The initial base bias setting is obtained from a potentiometer across the collector supply.

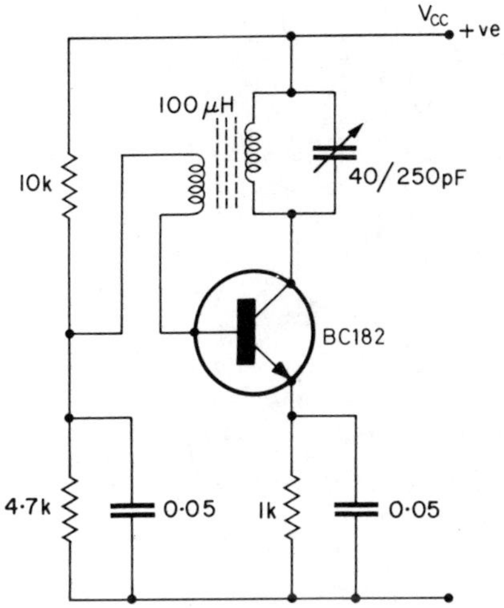

Fig. 11.1.—*Tuned collector feedback oscillator.*

The tuned collector feedback arrangement can also be applied to the common base circuit as shown in Fig. 11.2. It is very similar to the common emitter circuit, but the sense of the feedback must be arranged to be in phase with the collector since there is no phase change in the transistor. This arrangement is particularly useful for the self-oscillating mixer circuit commonly used for long and medium wave superheterodyne receivers. This will be described in a later chapter.

Since with the common base circuit the emitter and collector signals are almost in phase, the feedback from collector to emitter can be achieved by capacitive feedback as shown in Fig. 11.3. The r.f. choke L_2 in the emitter presents a high impedance at the frequency of oscillation, so that the feedback current flows into the emitter. This circuit is very useful at

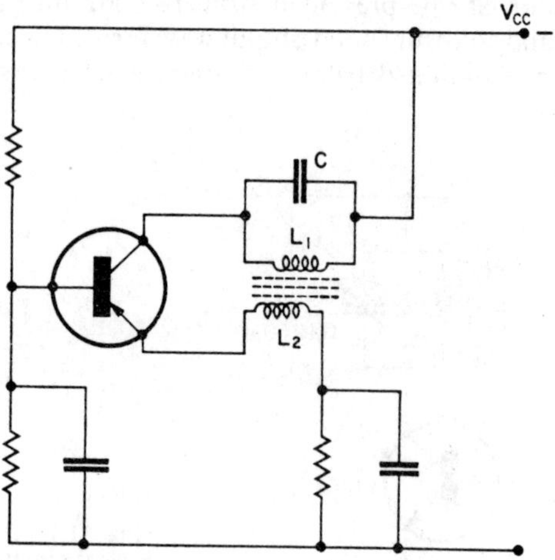

Fig. 11.2.—*Common base feedback oscillator.*

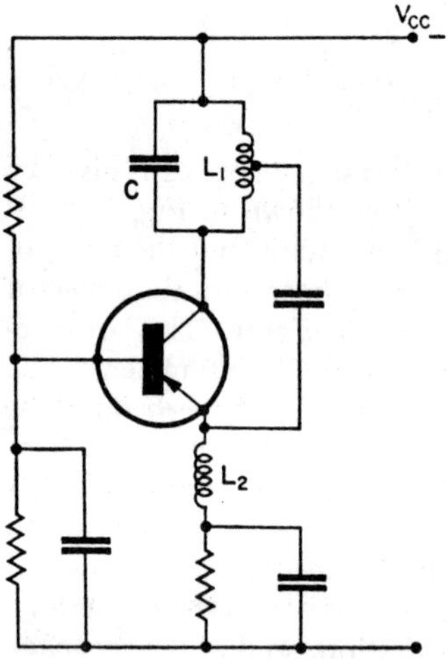

Fig. 11.3. — *Collector-emitter feedback oscillator.*

v.h.f. Fig. 11.4 shows a typical arrangement for a 250 Mc/s oscillator. The feedback capacitor is only 1 pF, and the 3,000 ohms resistor in the emitter presents a sufficiently high impedance.

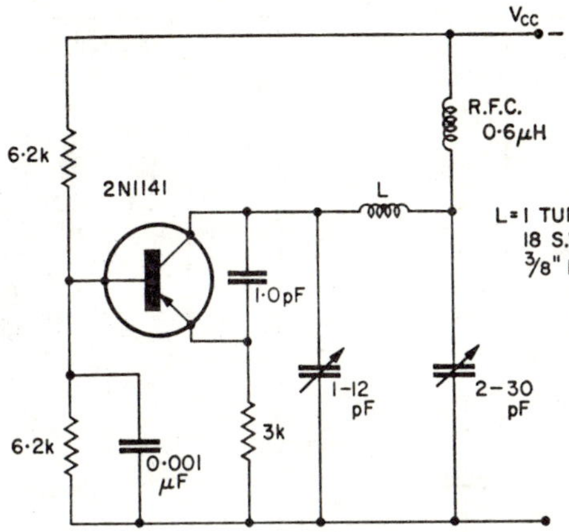

Fig. 11.4.—250 *Mc/s oscillator.*

At higher frequencies, in the u.h.f. band, it is possible to obtain oscillation with no external capacitance, the transistor internal collector to emitter capacitance being sufficient, but it is usual to design with a small external capacitance. For this band, it is customary to use quarter-wave or half-wave lines as the resonant circuits. A typical u.h.f. oscillator circuit is included in the u.h.f. television tuner (Fig. 10.13) described in the previous chapter.

Hartley and Colpitts oscillators

The classical Hartley and Colpitts circuits can be applied to common emitter transistor oscillators as shown in Fig. 11.5 and 11.6 respectively. The values in each case are suitable for a frequency of 100 kc/s. In the Hartley circuit, the feedback is obtained by means of a tapped inductance in the tuned circuit. The collector and base are connected to the opposite ends of the tuned circuit with the intermediate tap effectively connected to the emitter. In the Colpitts circuit, the emitter is connected to an intermediate point on the tuned circuit in the

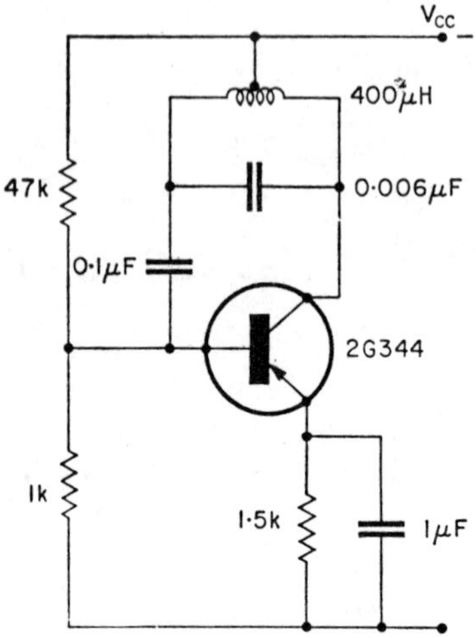

Fig. 11.5.—*Hartley oscillator.*

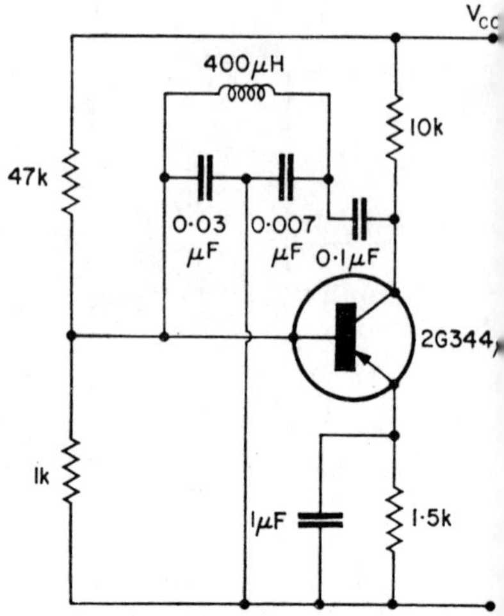

Fig. 11.6.—*Colpitts oscillator.*

capacitive arm by arranging the capacitance in a suitable series combination. In each oscillator bias stabilisation is used by means of a potentiometer across the supply, and a by-passed resistor is included in the emitter.

Maximum frequency of oscillation

The maximum frequency of oscillation depends entirely on the characteristics of the transistor. It was mentioned in the previous chapter that the power gain of a unilateralised amplifier at a frequency f is given by

$$A = \frac{\alpha_o f_T}{8\pi r_b C_c f^2}$$

The maximum frequency of oscillation, f_{max}, will be the frequency at which the power gain has fallen to unity. Then

$$f_{max} = \sqrt{\left(\frac{\alpha_0 f_T}{8\pi r_b C_c}\right)}$$

Table 10.1 includes typical values of f_{max} for the germanium alloy, the germanium alloy diffused, two types of silicon epitaxial planar and the germanium high frequency diffused base mesa transistor.

Microwave oscillator

Microwave transistor oscillators up to the order of 3 to 4 Gc/s are now possible with very high frequency silicon epitaxial planar and germanium epitaxial mesa transistors. Fig. 11.7 shows a typical arrangement using a TIX3016 silicon

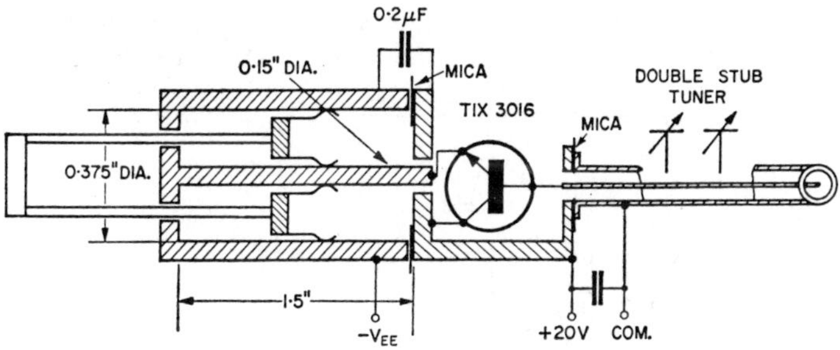

Fig. 11.7.—2,000 *Mc/s oscillator.*

transistor. The tuned cavity and concentric line components follow normal microwave practice. The main coaxial cavity is connected across the collector to emitter path of the transistor, and internal collector to emitter feedback is used with a tuned line in the base circuit. With a collector voltage of 20 volts and collector current of 15 mA an output of 50 mW can be obtained at 2 Gc/s.

Frequency stability

It is important in any oscillator to ensure that the oscillation frequency remains constant against changes of load impedance, supply voltage and ambient temperature. With tuned oscillators the Q of the tuned circuit should be kept high by using low loss inductances and by keeping the C to L ratio of the tuned circuit high. This reduces the magnitude of harmonics in the output and tends to stabilise the frequency on account of the highly frequency selective characteristic. The feedback circuit should be loosely coupled to the tuned circuit in order to minimise its effect on the circuit Q and to give better isolation from changes in the transistor characteristics. The d.c. operating point must be stabilised as described for low frequency amplifiers, since variations of the operating point may affect the output frequency and also the output waveform and amplitude.

Crystal control

To obtain an extremely stable fixed frequency, a quartz crystal can be used as a control element. The resonant frequency of a quartz crystal is determined purely by its natural

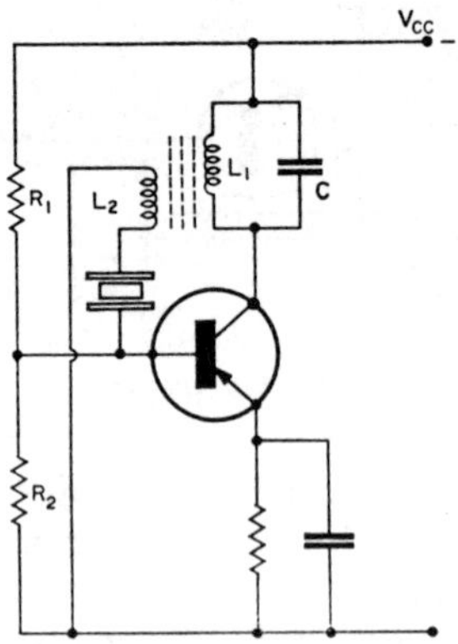

Fig. 11.8.—*Crystal controlled oscillator.*

mechanical vibrating characteristics and so is very stable. One method of applying the crystal is to include it in the feedback circuit as shown in Fig. 11.8. At frequencies above and below the series resonant frequency of the crystal, the impedance of the crystal rises and reduces the feedback. Thus the crystal only allows feedback at its critical resonant frequency, and so the oscillation frequency is very precise and stable. In the arrangement shown, the bias resistor R_2 is not decoupled since the feedback must be maintained at the base electrode.

Frequency multiplication

Quartz crystals are only available with fundamental resonant frequencies up to the order of 10–20 Mc/s. In order to obtain higher stable frequencies the frequency of the output from a crystal controlled oscillator can be multiplied by a suitable circuit arrangement. Frequency multiplication is usually accomplished by feeding the fundamental signal into a non-linear system, thus distorting the waveform and introducing harmonics. The required harmonic is selected by means of a tuned circuit in the output circuit. With transistors the emitter–base diode characteristic can conveniently be used as the non-linear system, and the fundamental signal is fed into the transistor with zero bias. The signal is distorted and amplified to the collector circuit, where a resonant circuit selects the required harmonic.

A typical frequency doubler circuit for 20 to 40 Mc/s is shown in Fig. 11.9. The common emitter circuit is used, and the input signal is matched to the transistor by means of a π-matching network. The second harmonic is developed across a tuned circuit, also arranged as a π circuit in order to match out into a 50 ohms load. A wave-trap circuit is included in the collector circuit to reject the fundamental component. In this wave-trap circuit C_1 and L_1 form a parallel tuned circuit at the second harmonic, thus presenting a high impedance which does not damp the main tuned circuit, but the combination of C_1, C_2 and L_1 forms a low impedance series tuned circuit at the fundamental frequency which shunts the output.

Frequency multiplying circuits are also used to obtain frequencies above the maximum frequency of oscillation of the available transistor. Fig. 11.10 shows a common emitter circuit to double from 120 to 240 Mc/s. The operation of the circuit is similar to the lower frequency circuit described above.

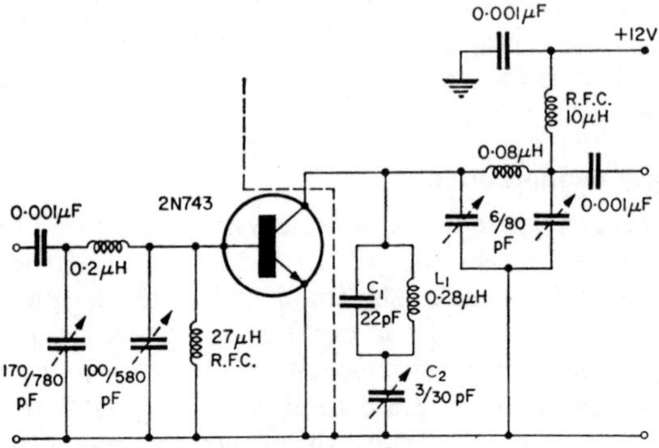

Fig. 11.9.—20–40 *Mc/s frequency doubler.*

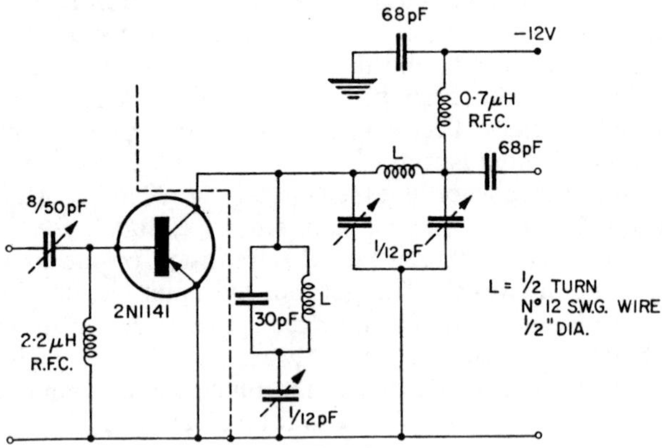

Fig. 11.10.—120–240 *Mc/s frequency doubler.*

Resistance–capacitance feedback oscillators

A sinewave oscillator can be obtained by using a resistance–capacitance circuit in place of the resonant system described above. The resistance–capacitance network is connected between the collector and base as a feedback network, with the transistor in the common emitter arrangement. Phase shift occurs in the *RC* network, and the circuit will oscillate at the frequency at which the phase shift is 180°. A typical arrangement is shown in Fig. 11.11. It uses a resistance–capacitance

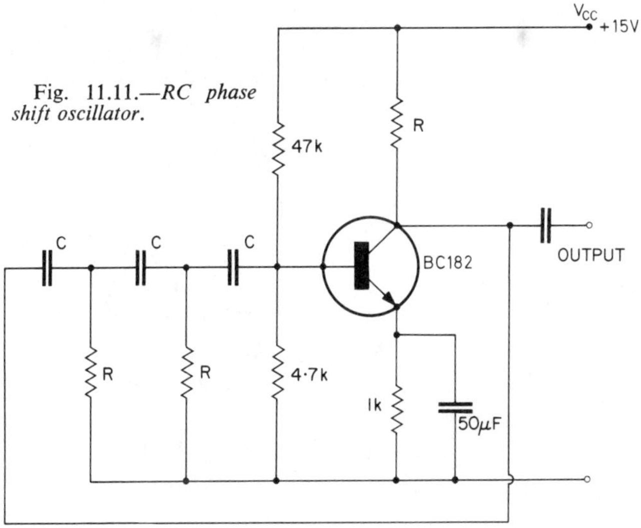

Fig. 11.11.—*RC phase shift oscillator.*

ladder with three steps, each step giving 60° phase shift at the frequency of oscillation. The main difference from the tuned *LC* oscillator is that the phase shift in the *RC* combination is accompanied by a substantial attenuation, and so it is usually necessary for the transistor to have a high gain. The value of *R* is chosen to be roughly the geometric mean of the transistor output and input impedances, a value in the region of 5,000 to 10,000 ohms being usual.

The approximate frequency of oscillation is given by $f = 0{\cdot}08/RC$. With $R = 6{,}800$ ohms and $C = 0{\cdot}01\ \mu F$ the circuit will oscillate at approximately 1,200 c/s. For other frequencies the capacitor or resistor value should be adjusted accordingly.

This circuit arrangement is most useful for single frequency oscillators. It is not convenient for variable frequency use, since the components of each step of the ladder would have to be varied simultaneously.

Wien bridge oscillator

An alternative oscillator circuit using a resistance–capacitance feedback network is the Wien bridge oscillator shown in Fig. 11.12. The circuit uses two transistors in the

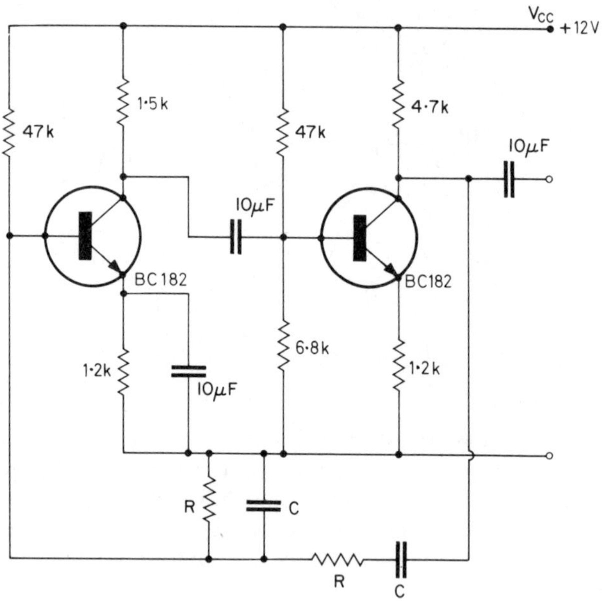

Fig. 11.12.—*Wien bridge oscillator.*

common emitter connection. The output from the second stage is fed through a series *RC* combination to a parallel *RC* circuit using the same component values. The voltage produced across the parallel circuit has zero phase shift at only one frequency, and so if this voltage is fed back to the input of the first stage positive feedback will occur at the one particular frequency, resulting in oscillation. The frequency of this oscillation is approximately equal to $0{\cdot}16/RC$. With $R = 10{,}000$ ohms and $C = 0{\cdot}01\ \mu F$ it will be 1,600 c/s.

For a variable frequency oscillator, the resistors R can be varied by means of ganged potentiometers, and the capacitors C can be switched for range changing.

RELAXATION OSCILLATORS

An oscillator circuit which is designed to give a non-sinusoidal output waveform is called a relaxation oscillator. Such oscillators generally use regenerative feedback together with a resistance–capacitance or resistance–inductance combination to give a time dependent variation. The charge and discharge times of the capacitance or inductance are used to give sawtooth, square or pulse output waveforms. Relaxation oscillators can be designed to be triggered by external synchronising impulses in order to "lock" the oscillation frequency to any desired signal. Typical relaxation oscillators are multivibrators, blocking oscillators and sawtooth generators.

Multivibrator circuits

The multivibrator is a two stage cross-coupled circuit in which at any given time one stage is conducting and the other is cut off. These conditions persist for a duration determined by the time constant of an RC combination, after which the stage which was cut off becomes conducting and in doing so causes the other to become cut off. The basic transistor multivibrator circuit is shown in Fig. 11.13 (a). The two transistors will not be exactly matched in gain and so when the circuit is switched on the current through the higher gain transistor, say T_1, will increase faster than that through the other. The collector voltage of T_1 will fall, and the change will be fed back to the base of transistor T_2, and so reduce its collector current. This will cause the collector voltage of T_2 to increase, and this increase will be fed to the base of T_1 and increase its collector current further. This regenerative action will continue until

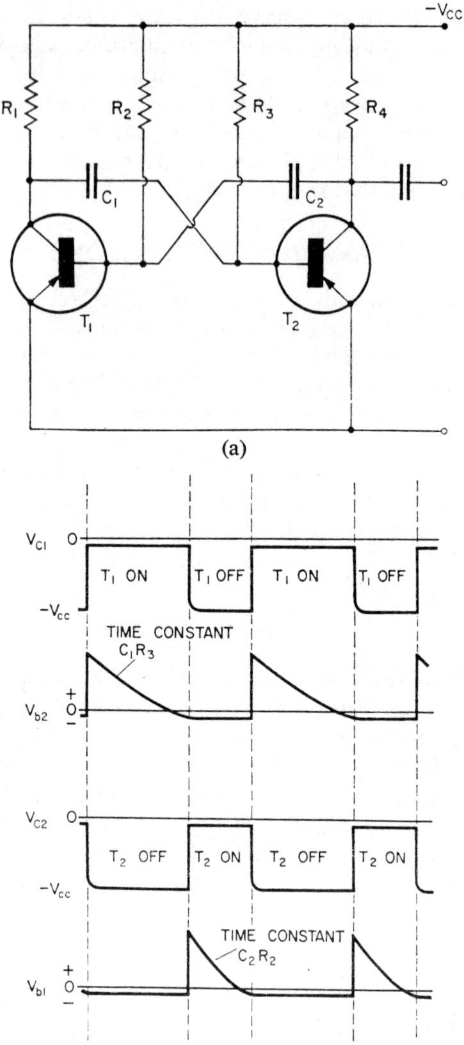

Fig. 11.13.—(a) *Basic astable multivibrator circuit.* (b) *Voltage waveforms of astable multivibrator.*

T_1 is completely on and T_2 is cut off. Capacitor C_1 now charges up through R_3 until the base of T_2 becomes sufficiently negative that T_2 starts to conduct. Then T_2 turns on and T_1 cuts off; C_2 charges up through R_2 and the cycle repeats. Fig. 11.13 (b) shows the voltage waveforms at the two collector and base electrodes.

Considering transistor T_1 we have

$$i_c = \beta i_b$$

and

$$i_b \doteqdot \frac{V_{CC}}{R_2}$$

At current saturation, during the ON state,

$$i_c \doteqdot \frac{V_{CC}}{R_1}$$

Thus

$$\frac{V_{CC}}{R_1} \doteqdot \beta \frac{V_{CC}}{R_2}$$

i.e.

$$\frac{R_2}{R_1} = \beta$$

Thus to ensure that the transistor saturates the current gain β must be greater than the ratio R_2/R_1. The value of R_1 is a compromise. It should be kept low enough to avoid effects due to variation of cut-off current, but must not be so low as to reduce the stage gain materially.

Typical values are $R_1 = 1{,}000$ ohms and $R_2/R_1 < 20$. The time for alternate on and off periods will be approximately $0{\cdot}7C_1R_3$ and $0{\cdot}7C_2R_2$. In a symmetrical circuit, with C_1 equal to C_2 and R_2 equal to R_3, a square waveform is obtained with equal ON and OFF times, and a frequency of $0{\cdot}7/CR$. With $R_2 = R_3 = 10{,}000$ ohms and $C_1 = C_2 = 0{\cdot}007\,\mu$F the oscillation frequency will be approximately 10,000 c/s.

A practical circuit useful to demonstrate the action of a free-running multivibrator is shown in Fig. 11.14. The circuit switches between the two transistors at a frequency around

1 c/s. As each transistor switches into the conducting state, the corresponding lamp in its collector circuit lights, and so each lamp is on for half a second.

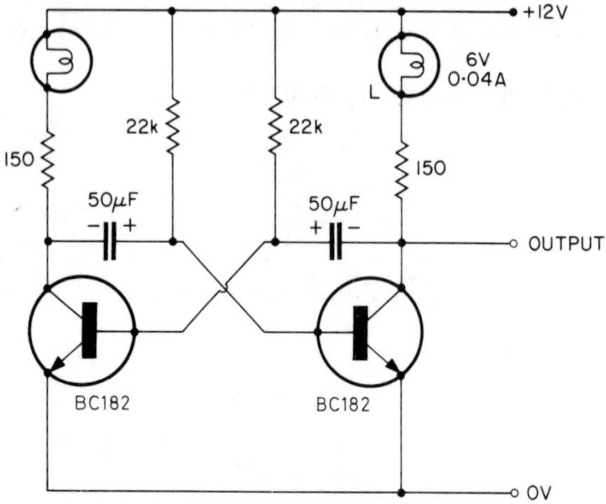

Fig. 11.14.—*Practical low-speed astable multivibrator circuit.*

The basic multivibrator circuit described above can be modified to give two other useful modes of operation, monostable and bistable. These illustrate important concepts in switching circuits and are discussed in Chapter 13.

Blocking oscillator

In a blocking oscillator it is arranged that the transistor conducts for a short period, during which time a capacitor charges up. Then the circuit drives the transistor to cut-off, and the capacitor discharges slowly through a resistor. A typical circuit is shown in Fig. 11.15. When the supply voltage is switched on base current flows into the transistor via resistor *R*, causing collector current to flow through the transformer primary. Feedback regeneration to the base circuit takes place, the transistor is rapidly driven ON into saturation and capacitor *C* charges up. At saturation, the collector current becomes constant and regeneration ceases. The field in the transformer collapses, and a reverse voltage is induced in the base winding which drives the base to cut-off and turns the transistor OFF. The voltage across the capacitor C is such as to initially hold the transistor in cut-off, but it gradually discharges through the resistance R until the voltage has fallen

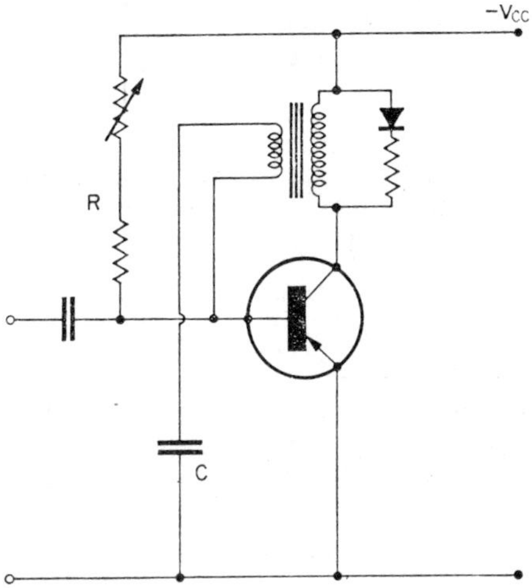

Fig. 11.15.—*Blocking oscillator circuit.*

to the point where the transistor starts to conduct again. Then the cycle repeats. The diode and resistor across the collector winding of the transformer prevent excessive voltage when the field collapses.

The main use of the blocking oscillator is as a pulse generator, the output being taken from the collector.

Sawtooth waveform generators

Sawtooth voltages are used for various applications, the most common being the scanning of cathode-ray tubes and the linear variation of voltage or frequency for display purposes.

Unijunction transistor oscillator

Perhaps the simplest of all oscillator circuits is that using the unijunction transistor. The circuit is shown in Fig. 11.16.

The capacitor *C* charges up slowly through the resistance *R* until the emitter voltage reaches the value at which the transistor triggers into conduction. Then the capacitor discharges

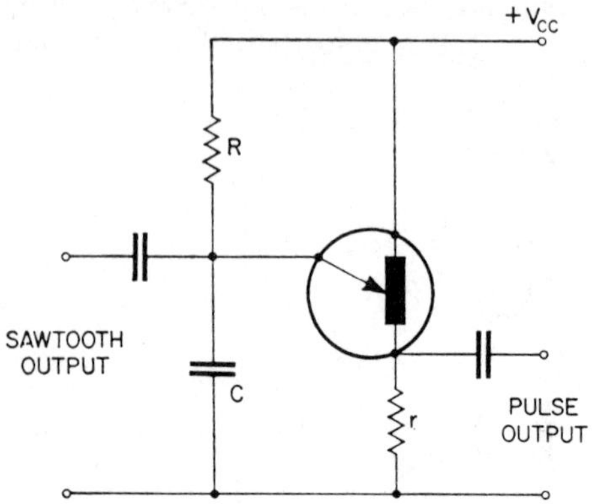

Fig. 11.16.—*Unijunction transistor oscillator.*

quickly through the transistor and the small resistance *r*. When the voltage has fallen to a low value, the transistor returns to the non-conducting state and the capacitor starts to charge up again. Thus a sawtooth voltage is produced across the capacitor *C*, and short pulses are developed across the resistance *r*.

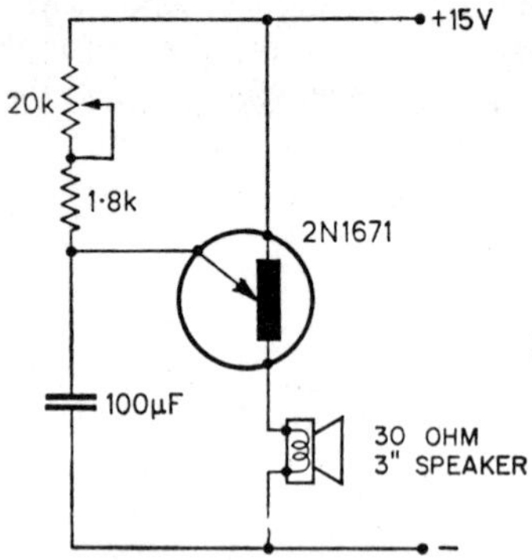

Fig. 11.17.—*Unijunction transistor metronome.*

Fig. 11.17 shows a practical unijunction transistor oscillator which might be called an electronic metronome. The capacitor discharge current pulses are fed through a small 30 ohms moving-coil loudspeaker and so produce audible "ticks". The frequency of the ticks can be controlled by varying the charging resistor. Apart from its obvious use in the musical world, such a unit can find application in the factory when short manually executed processes are carried out—the operator can time process durations by listening to the ticks.

Feedback sawtooth generator

A feedback circuit to produce a sawtooth waveform is shown in Fig. 11.18. On switching the supply voltage ON, the capacitor C charges up rapidly through the transistor T_2, and

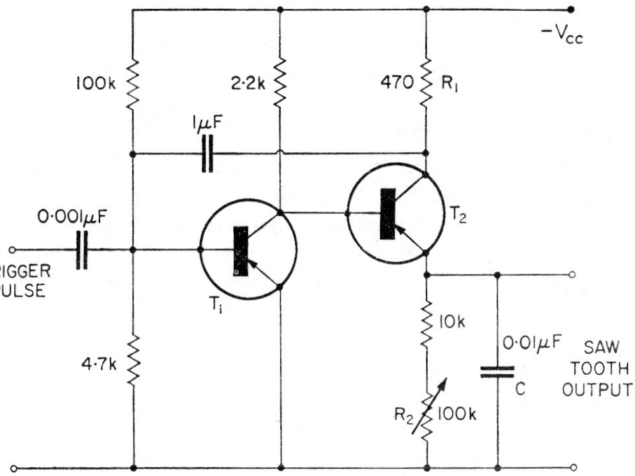

Fig. 11.18.—*Feedback sawtooth generator.*

at the same time the voltage pulse produced across the resistance R_1 is fed back to the base of transistor T_1 and cuts it off. As the charging current of C decreases, the pulse across R_1 decreases and allows T_1 to conduct, which causes T_2 to be cut off. Then C discharges slowly through R_2 until T_2 starts to conduct again.

12
TRANSISTOR RADIO RECEIVERS

THE various circuits described in the preceding chapters are combined in the design of complete radio receivers. Transistor receivers for domestic use are available in countless types, all with excellent performance and reasonable cost. It is not intended in this book to go too deeply into the design of commercial receivers, but rather to encourage the young engineer to feel his way into the receiver field by first describing several simple circuits and then outlining a typical complete superheterodyne receiver circuit. Finally one or two special receiver circuits will be mentioned.

Simple detector receivers

The simplest possible transistor receiver is shown in Fig. 12.1. For reasonable reception with this circuit it is essential that a good external aerial is used, that this is sited as high as possible and is at least 50 feet long. Also it is necessary to make a very low resistance earth connection, for example to a rising water main.

The aerial is fed to a series tuned circuit and the coupling from the inductance to the transistor is arranged for optimum power transfer by tapping down the inductance to match the low input impedance of the transistor to the dynamic impedance of the tuned circuit. A step down of the order of five to one is typical. Some adjustment of the input impedance of the transistor can be effected by varying the bias resistor to change the emitter current of the transistor. The emitter-base diode demodulates the signal and an amplified output is obtained in the collector circuit.

The effective tuning capacitance will be the capacitor *C* in series with the capacitance to earth of the aerial. Thus the actual value of capacitor *C* will depend upon the aerial installation. A 500 pF variable capacitor is convenient. For the

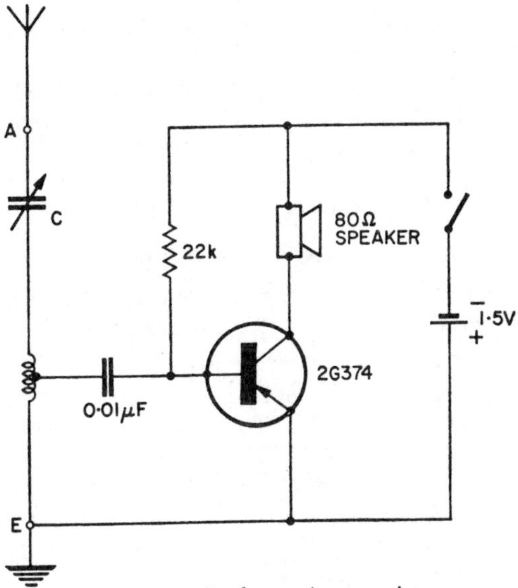

Fig. 12.1.—*Single transistor receiver.*

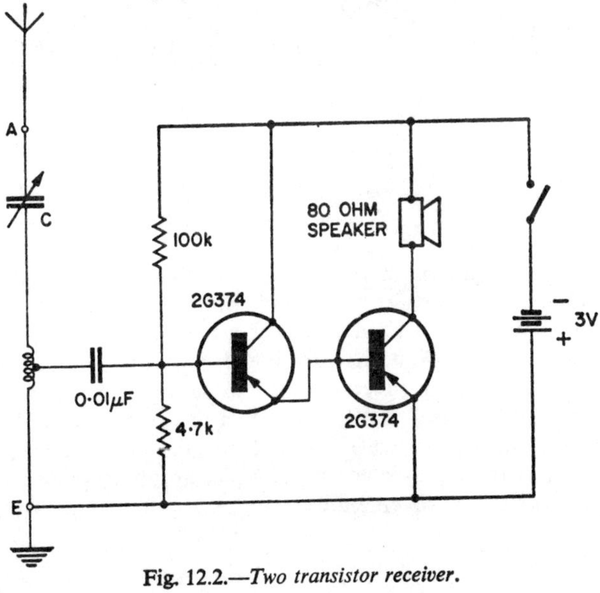

Fig. 12.2.—*Two transistor receiver.*

medium wave band, an inductance of 200 μH is satisfactory and for the long waveband 1 mH.

Additional amplification with a minimum of components can be obtained by means of a direct coupled amplifier stage as shown in Fig. 12.2. The emitter current of the detector transistor is fed direct to the base of an output transistor, which feeds the loudspeaker.

Regenerative detector receiver

Additional sensitivity and selectivity can be obtained by using regeneration in the detector stage. Fig. 12.3 shows a

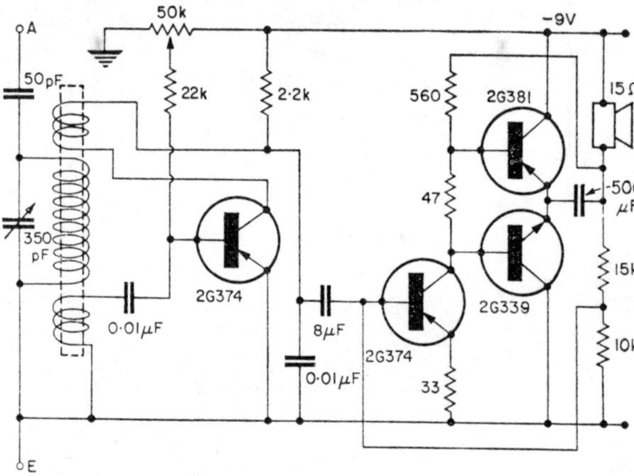

Fig. 12.3.—*Transistor regenerative receiver.*

receiver with regeneration and a two-stage audio amplifier suitable for local station reception. Emitter-base diode detection is used as in the simple detector receiver, and regeneration is included by feeding back from the collector to a winding coupled to the main tuned circuit. Variation of the base bias voltage will control both the detection efficiency and the regeneration, and so will form a convenient output control. The output from the detector is coupled to an amplifier stage which drives a complementary pair push-pull class B output stage. Depending upon the location, it may be possible to obtain sufficient output with the ferrite-rod aerial only, but an outside aerial can be fed in through a small capacitor.

Superheterodyne radio receiver

The circuit of a typical transistor portable receiver for the long and medium wavebands is shown in Fig. 12.4. It uses seven transistors, and gives a maximum audio output power of 0·5 watt. The sensitivity on the medium waveband is 200 μV per metre and on the long waveband 600 μV per metre.

The aerial consists of a ferrite rod on which the input circuit inductance and coupling coils are positioned. The mixer circuit is of the self-oscillating type with the input signal from the aerial coupling coil fed to the base of the mixer transistor. Oscillator feedback from collector to emitter is arranged by coupling both circuits to the main oscillator tuned circuit. The stage is d.c. biased by the base voltage potentiometer method.

The intermediate frequency signal is developed across the primary of the first i.f. transformer in the collector circuit of the mixer transistor. This feeds a two-stage 465 kc/s amplifier with both stages neutralised by means of feedback capacitors. Single tuned i.f. transformers are used and give satisfactory selectivity for the portable type of receiver. The output from the i.f. amplifier is connected to a conventional diode detector. The d.c. voltage developed across the detector load resistance is arranged to give reverse a.g.c. to the first i.f. stage. The second i.f. stage has fixed bias.

The detector load is a potentiometer which forms the volume control, and the slider is connected to an *RC* coupled amplifier stage which is followed by a driver and a complementary pair class B output stage. The audio output is coupled to the 15 ohms loudspeaker via an isolating capacitor. Negative feedback is applied from the output to the base of the driver transistor.

The sensitivity of the audio amplifier is 8 mV at the slider of the volume control to give 50 mW output and the sensitivity of the i.f. plus audio amplifiers is 4 μV at the base of the mixer

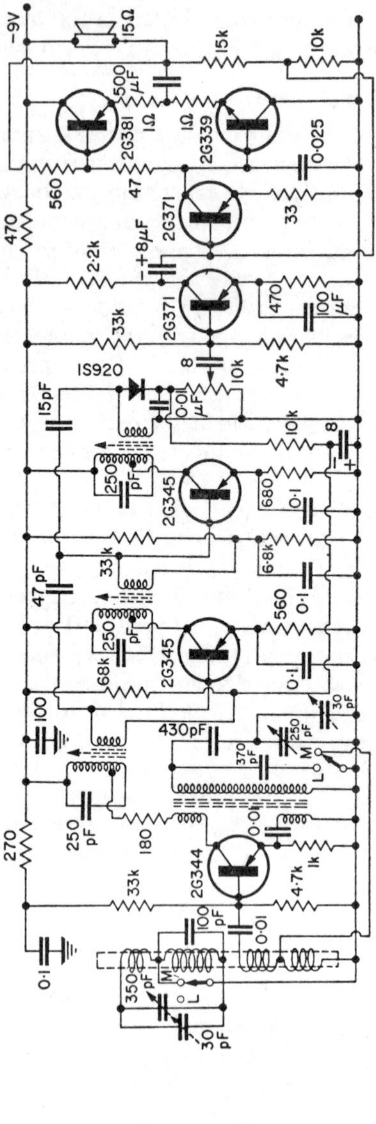

Fig. 12.4.—*Transistor superheterodyne receiver.*

transistor for 50 mW audio output. The supply voltage is 9 volts with a current consumption of 10 mA quiescent and 40 mA for 50 mW output.

Superheterodyne reflex receiver

In a reflex circuit a single stage is arranged to amplify a radio frequency signal and an audio frequency signal simultaneously. A block diagram showing how this can be carried out with a superheterodyne receiver is shown in Fig. 12.5. The incoming

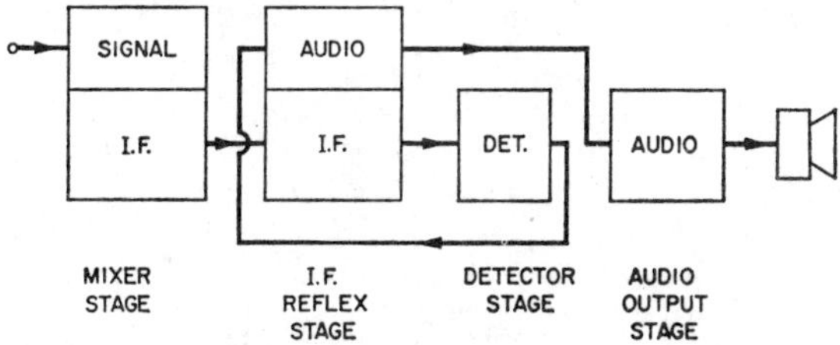

Fig. 12.5.—*Block diagram of a reflex receiver.*

signal is fed to the mixer and the i.f. output amplified as usual. After detection the audio signal is returned to the base circuit of the last i.f. amplifier stage and an amplified audio signal is developed across an audio load resistance included at the supply end of the collector circuit. This is then fed to the audio amplifier.

The obvious advantage of the system is that one reflex stage does the work of two normal stages, resulting in an overall saving of cost and space. The main disadvantage is that the circuit requirements are more precise and the design is less flexible.

A complete three transistor reflex receiver for the medium waveband is shown in Fig. 12.6. The aerial and mixer stages are similar to those for the normal superheterodyne receiver and the i.f. stage is also basically similar except that an audio load resistor R_3 is included in the collector circuit. The i.f. signal is fed to the diode detector, and the audio output fed back via the capacitor C_1 to the base bias point of transistor T_2.

This point is decoupled to earth for 465 kc/s by C_2, but this capacitor is kept small so that it does not affect the audio signal appearing across R_2. Since the secondary winding of the transformer $I.F.T_1$ is only an ohm or so at audio frequencies,

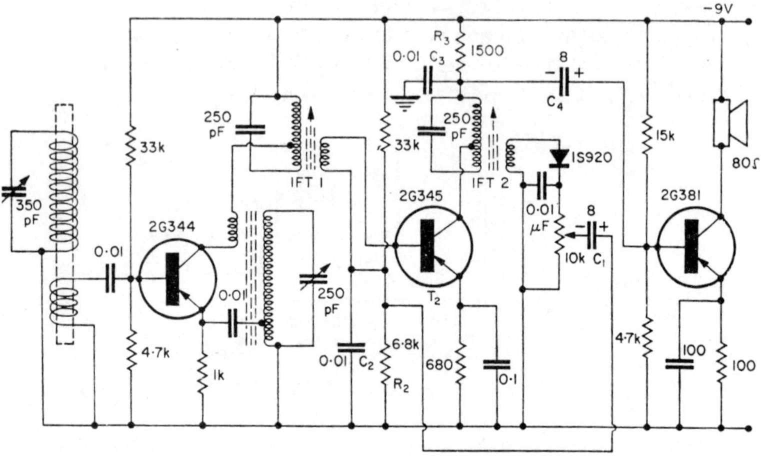

Fig. 12.6.—*Superheterodyne reflex receiver.*

the audio signal will be applied to the transistor base electrode in addition to the i.f. signal. In the collector circuit, the i.f. output is taken through $I.F.T_2$ and the circuit decoupled to earth for 465 kc/s by C_3. The audio output is developed across R_3 and is fed via C_4 to the base of the class A audio output transistor.

Super-regenerative receiver

The super-regenerative receiver is well known for its simplicity and excellent sensitivity, especially for high frequency use. Regeneration is used with a circuit arrangement which allows a build up to the oscillation point followed by a quenched period, this cycle being repeated at a supersonic rate. Fig. 12.7 shows a super-regenerative receiver for the

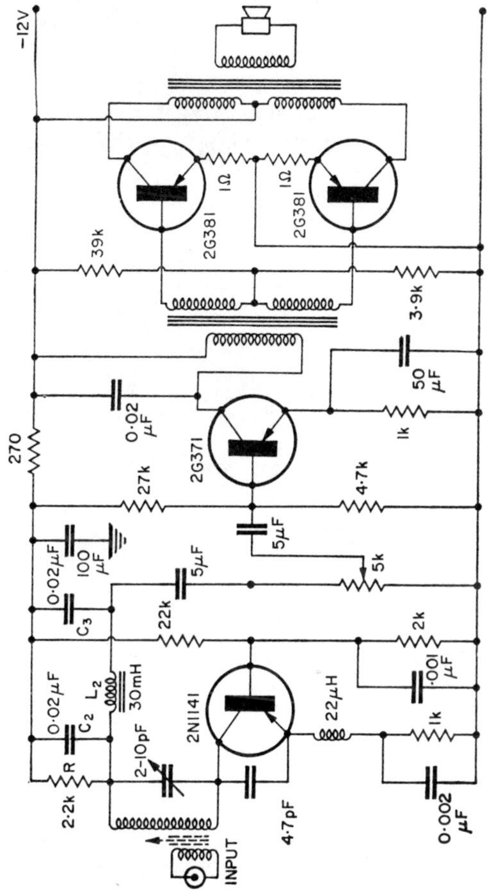

Fig. 12.7.—*Super-regenerative receiver for* 110–140 *Mc/s. Loudspeaker impedance* 8 *ohms.*

110–140 Mc/s band. The first transistor operates as a self-quenching detector, the quench frequency being determined by the time constant of the base circuit. The audio output appears across the collector load resistance *R*, and the combination of C_2 L_2 and C_3 form a low pass filter to remove the quench frequency from the audio output. The audio amplifier can be any convenient arrangement. That shown is a pre-amplifier stage and a driver stage transformer coupled to a class B push–pull output stage feeding an 8 ohms loudspeaker.

13
THE TRANSISTOR AS A SWITCH

An important application of the transistor is to control the flow of current through a load in a manner similar to a switch. Fig. 13.1 (a) shows a circuit incorporating a conventional

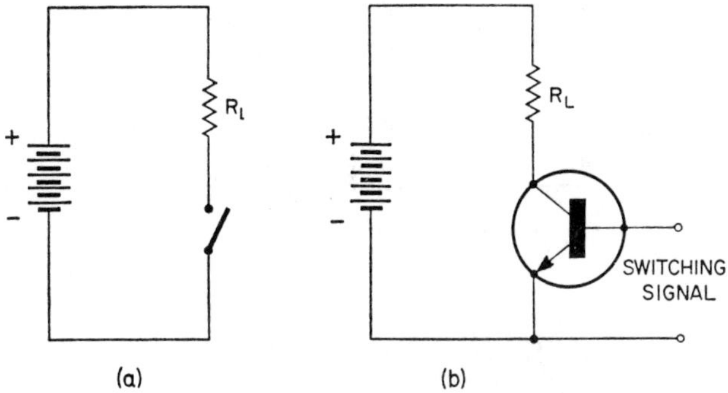

Fig. 13.1.—*The transistor used as a switch.*

switch with mechanically operated contacts. Fig. 13.1 (b) shows a corresponding arrangement in which the current through the load R_L is controlled by a transistor. With a reverse voltage applied to the base of the transistor, only the very small collector to base leakage current can flow through the transistor, and so there will be almost zero load current, the battery voltage appearing across the transistor. If the base voltage is made positive, the collector to emitter impedance will fall and load current can now flow through the transistor and load resistance, the voltage across the transistor falling to almost zero.

In practically all switching applications the transistor is used in the common emitter arrangement, since the high current gain allows large collector currents to be switched by means of a relatively small base current.

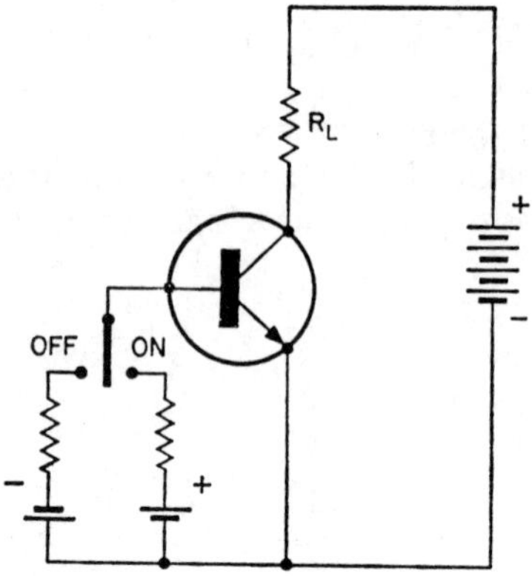

(a) Switching circuit

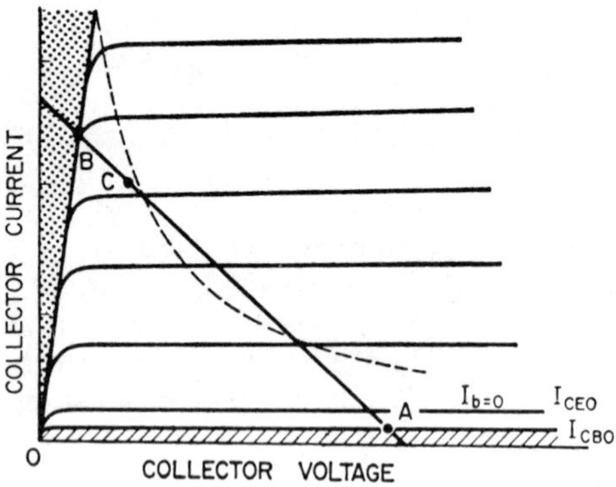

(b) Switching charcteristics

Fig. 13.2.—*Transistor switching characteristics.*

To study the characteristics of the transistor as a switch it is necessary to consider three modes of operation. In addition to the OFF state and the ON state, the transition region through which the transistor must pass in going from one state to the other is important. Consider the circuit of Fig. 13.2 (a) and the collector characteristics of Fig. 13.2 (b). The load line corresponding to the load resistance R_L is shown on the characteristics.

The OFF condition

With the base current at zero the collector current will be the collector-to-emitter leakage current I_{CEO}. If a reverse voltage is applied to the base, the collector current is reduced to the collector-to-base leakage current I_{CBO}. This current is very small—with silicon planar transistors of the order of 1 nanoampere, giving an OFF impedance above 1,000 MΩ. With germanium alloy transistors I_{CBO} is of the order of 1 μA, giving an OFF impedance above a megohm. When the base–emitter junction of a transistor is reverse biased, the transistor is said to be cut off, and is in the shaded region of Fig. 13.2 (b). With the load resistance R_L the operating point will be at A.

The ON condition

As the base current is increased, the collector current increases, and the operating point moves along the load line and eventually reaches point B. Then further increase of base current will not result in any more increase of collector current, and the transistor is said to be *saturated* or *bottomed.* The collector to emitter voltage will have fallen to a very low value, known as the collector saturation voltage, $V_{CE(\mathrm{sat})}$. In this condition the resistance between the collector and emitter of the transistor is very low—only a few ohms. As the collector voltage falls to the low $V_{CE(\mathrm{sat})}$, the collector junction becomes forward biased. Thus in the ON condition both junctions of the transistor are forward biased, and the current flowing through the load resistance will be $(V_{CC}-V_{CE(\mathrm{sat})})/R_L$, or approximately V_{CC}/R_L which is independent of the base current. The saturation region is shown dotted in Fig. 13.2 (b).

The transition region

Fig. 13.3 illustrates what happens during the transition from the OFF state to the ON state. As the base current is increased the collector current increases and the collector voltage falls. In the OFF state the collector dissipation is very low. With a germanium transistor at an ambient temperature of about 55°C the leakage current may be about 30 μA, and with a V_{CE} of 12 volts the dissipation will be 0·36 mW, which is quite negligible.

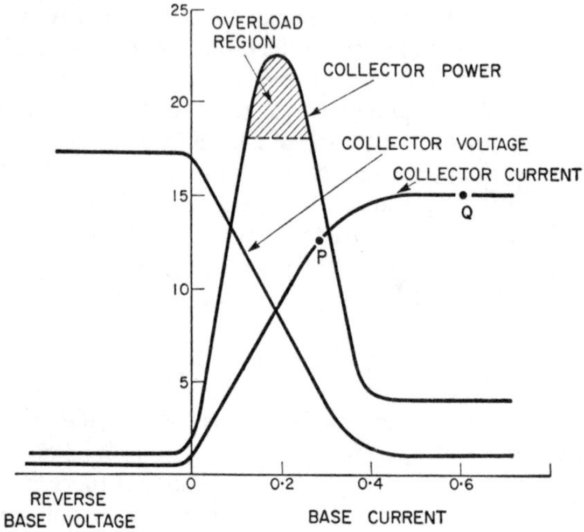

Fig. 13.3.—*Transistor characteristics.*

In the ON state also the power dissipation is low. The above germanium alloy transistor will have a $V_{CE(sat)}$ of about 0·2 volt at a current of 100 mA, giving a dissipation of 20 mW —again negligible.

As the collector current increases from the OFF state the transistor dissipation increases as shown in Fig. 13.3. It passes through a maximum point and falls to a low value again as saturation is reached. It is important to note that at its maximum value the power dissipation may be in excess of the rating of the transistor, and it would not be possible to operate the transistor permanently in that condition. To ensure the most satisfactory operation from the power dissipation viewpoint the base signal must be sufficient to drive the transistor into saturation very quickly, so that the mean power dissipation is kept low. Referring to Fig. 13.2 the base current should be larger than that needed to take the operating point to B to ensure that the transistor is driven into saturation. The effective current gain will then be lower than the small signal current gain h_{fe}. At the point P (Fig. 13.3), in the active transition region, the small signal gain may be about 40. If the base current is increased to the point Q to ensure saturation the effective gain will only be about 25.

For some applications it may be desirable to arrange that the transistor does not go into saturation, that is, so that the operating point only moves to the point P in Fig. 13.3 or the point C in Fig. 13.2 (b). This can be arranged either by limiting the value of the base current drive, or by clamping (which will be described later). The arrangement is then called a non-saturating switch, and care must be taken that the power rating of the transistor is not exceeded.

Pulse response

Given satisfactory OFF and ON states, the performance of a transistor switch depends upon its response to a short duration impulse. Suppose a transistor is OFF and a rectangular base current pulse as shown in Fig. 13.4 is applied. The shape of the collector current pulse is shown. It can be divided into five distinct regions, the delay time, t_d, the rise time, t_r, the pulse time, t_p, the storage time t_s and the fall time t_f. Consider these regions in turn. When the base current increases, holes (pnp transistor) are injected from the emitter into the base region, and a finite time is taken for these holes to travel across to the collector junction. This is the *delay time* and is defined as the time to the point where the collector current is 10 per cent of the final value. Owing to the dispersion of the holes in passing through the base region a further time passes before they have all reached the collector. This is the *rise time* and is that between the 10 per cent and 90 per cent points. The *pulse time* depends upon the duration of the input pulse. During the pulse time, current is flowing through the transistor, and holes are spread throughout the base region. When the base current is returned to zero, holes cease being

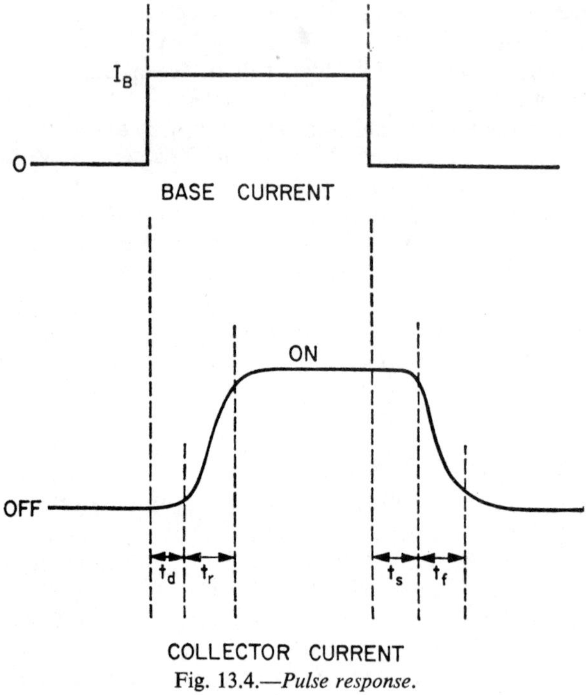

Fig. 13.4.—*Pulse response.*

injected from the emitter, but those already present in the base region continue to flow and maintain the collector current for a time until they all cross the collector junction and recombine. The time from the end of the base pulse to the point where the collector current has fallen to 90 per cent of its value is called the *storage time*, being due to the "storage" of holes. Finally the collector current will gradually fall as the holes recombine.

The time between the 90 per cent and 10 per cent points is called the *fall time*.

The more carriers stored in the base region the longer will be the storage time. When the collector current is saturated,

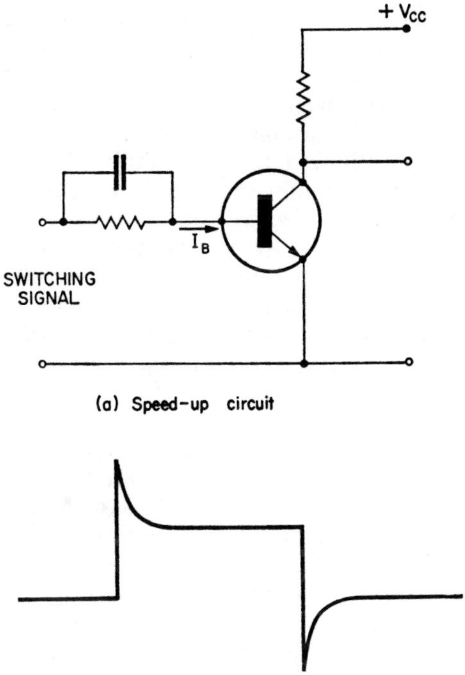

(b) Base current waveform

Fig. 13.5.—*Use of a speed-up capacitor.*

the collector voltage falls below the base voltage, and so the collector junction becomes forward biased and injects more carriers into the base region. Thus with saturated switches the storage time is longer than with non-saturated switches, and for this reason the latter are preferred for high speed switching applications.

The *switch-on-time*, t_{ON}, is equal to the sum of the delay and rise times, $t_d + t_r$, and the *switch-off-time* the sum of the storage time and the fall time $t_s + t_f$. The sum of the switch-on and switch-off times is called the *total switching time*, t_T, and is used as a figure of merit for switching transistors. With high speed silicon epitaxial planar transistors, such as the 2N743, the total switching time for saturated switching is about 30 nanoseconds and for non-saturated operation about 10 nanoseconds. With the lower speed germanium alloy switching transistors the saturated total switching time is about 1 μsec.

In practical switching circuits it would be ideal to overdrive the transistor heavily during the turn-on time, to obtain a fast rise, then to drop the drive to just saturation level to reduce the storage time, and finally to overdrive the transistor with a large reverse pulse to turn it off. The circuit shown in Fig. 13.5 (a) gives these conditions, producing a base current with a waveform as indicated in Fig. 13.5 (b). The capacitor, called a *speed-up-capacitor* provides the initial and end spikes.

Diode clamping

The ON condition of a transistor can be well defined in the saturated state by means of suitable overdrive to ensure that saturation is reached under all conditions. If non-saturated operation is required, it is very difficult to define the ON condition by close control of the amplitude of the base current, and a more satisfactory method is to clamp the collector voltage at the required level by means of a suitably biased diode. The method is illustrated in Fig. 13.6. Assume that the transistor is cut off at the point A. If the base current is increased the collector current increases and the collector voltage falls as the operating point moves along the load line. When the collector voltage has fallen to a value equal to the diode bias voltage E_d, the diode will start to conduct and the collector will be held at the potential E_d. Additional collector current will now be drawn from the diode bias supply but, since this current does not pass through the load resistance, the voltage at the collector remains approximately at E_d and so the transistor does not go into saturation.

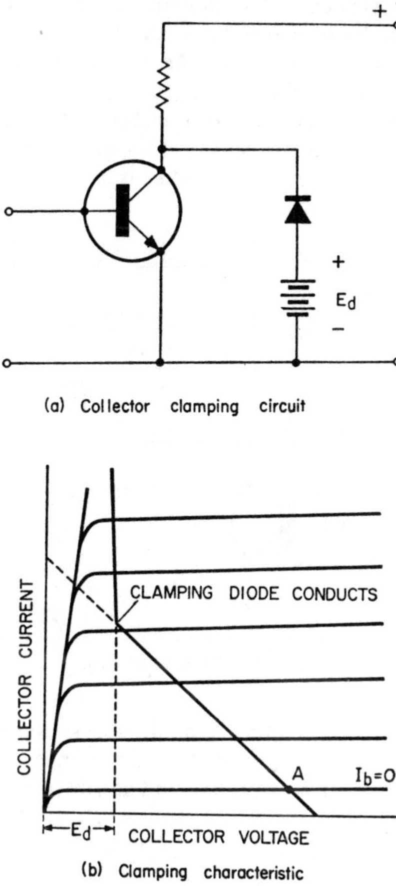

Fig. 13.6.—*Collector saturation clamping.*

Relay operation

In various industrial applications it may be required to operate a relay or contactor by energising its operating coil from a transistor. Since there will be inductance present in the transistor load certain precautions must be taken. The transistor will normally be driven into saturation, and the d.c. collector supply voltage must not be greater than the BV_{CEO} rating. At turn-off the decay of energy stored in the inductance induces a voltage which adds to the supply voltage to form a collector to emitter voltage spike. This spike could exceed the voltage rating of the transistor, and so it is necessary to protect

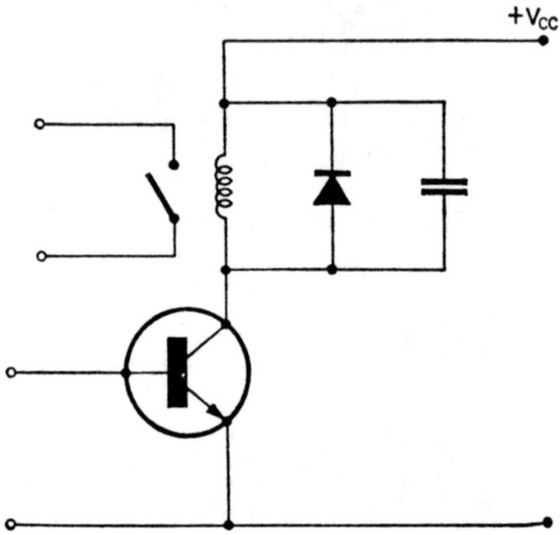

Fig. 13.7.—*Relay operation.*

the transistor by connecting a diode across the relay coil as shown in Fig. 13.7. The capacitor, also connected across the coil, gives a more linear load line and assists in suppressing the voltage spikes.

TRIGGERED SWITCHING CIRCUITS

Switching circuits are usually arranged to be stable in either the cut-off or saturation state, and are triggered into the other by means of an applied pulse. The pulse moves the operating point of one of the transistors into the transition region; the transistor becomes active, with a current gain, and regeneration occurs to complete the change over. The most common examples of this type of circuit are the monostable and bistable versions of the multivibrator circuit described in Chapter 11. The regenerative action in both circuits is similar to that previously described for the basic circuit.

Monostable circuit

In the monostable arrangement the circuit is designed to be quiescent in the required state. An external impulse will

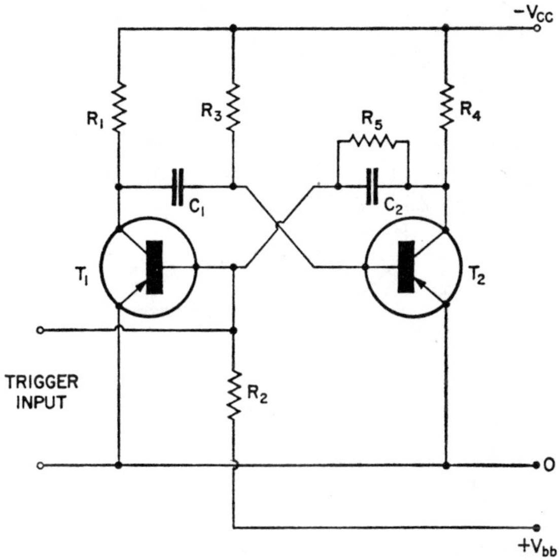

Fig. 13.8 (a).—*Basic monostable multivibrator circuit.*

trigger it into the other state and after a period of time determined by the time constant of a resistance–capacitance combination the circuit automatically reverts to the original state and becomes quiescent again until a further impulse is applied. The arrangement is shown in Fig. 13.8 (a). By means of the base voltage V_{bb} the transistor T_1 is biased to cut-off, which results in T_2 being in the ON state. These are the quiescent conditions. If a negative impulse is applied to the base of T_1, its operating point will move into the active region. Immediately regeneration will take over, and the circuit will trigger T_1 ON, into

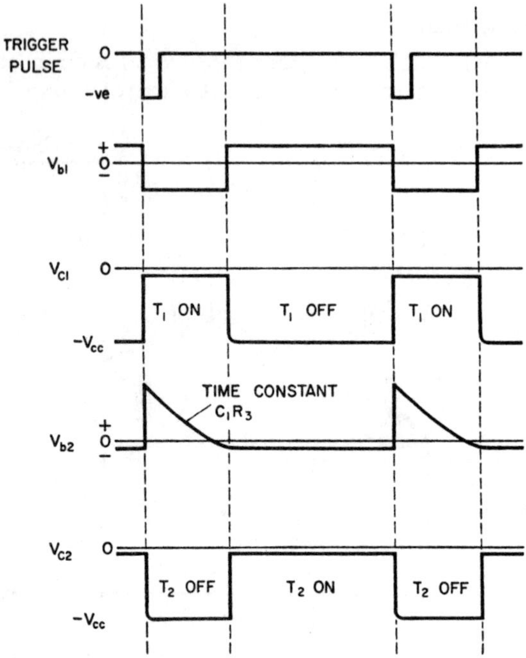

Fig. 13.8 (b).—*Monostable multivibrator voltage waveforms.*

saturation, and T_2 OFF, into cut-off. Then the capacitor C_1 charges through R_3 until T_2 starts to conduct again and the circuit triggers back to its original state. The circuit gives one output pulse for each trigger pulse, as shown in Fig. 13.8 (b). The duration of the output pulse is decided by choice of the time constant of $C_1 R_3$.

Bistable circuit

A bistable circuit can remain stable in either of the two states. A first impulse triggers the circuit from one state to the other, and it remains stable in that state until a second impulse is applied, when the circuit triggers back to the first state. The circuit is shown in Fig. 13.9 (a). If T_1 is conducting, its collector potential will be near zero; T_2 will have a positive base bias and so will be cut off. This means that the collector of T_2 will be at the full negative supply potential and so the base of T_1 will be negative, holding T_1 in saturation. If a negative impulse is

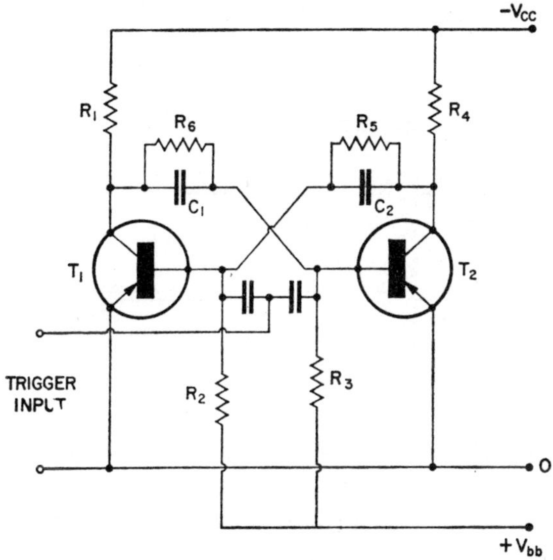

Fig. 13.9 (a).—*Basic bistable multivibrator circuit.*

applied to the base of T_2 its operating point moves into the active region and the circuit triggers T_2 ON and T_1 OFF. In this state the bias conditions of the two transistors are reversed, and the circuit is stable in the new condition. Now, a negative pulse on the base of T_1 will trigger the circuit back to the first state, and so on. Thus each transistor is switched ON once for every two trigger pulses. The voltage waveforms at the two collectors are shown in Fig. 13.9 (b).

When triggering the circuit on and off repeatedly with input pulses of the same polarity it is desirable to include a pulse steering circuit in the input so that the input pulse is directed to the proper transistor only. A convenient steering circuit is shown in Fig. 13.10. To describe the operation of the circuit

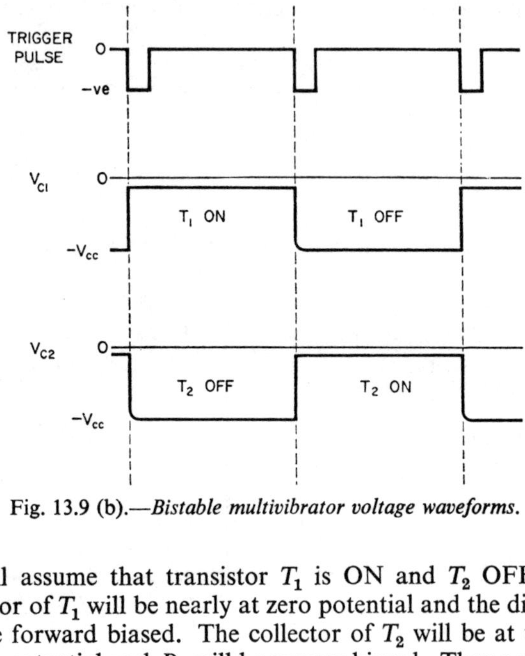

Fig. 13.9 (b).—*Bistable multivibrator voltage waveforms.*

we will assume that transistor T_1 is ON and T_2 OFF. The collector of T_1 will be nearly at zero potential and the diode D_1 will be forward biased. The collector of T_2 will be at the full supply potential and D_2 will be reverse biased. Thus a positive input trigger pulse will be "steered" through D_1 rather than through D_2 and transistor T_1 will be turned OFF, resulting in T_2 coming ON. The bias conditions will then be reversed, and the next pulse will be steered through D_2 to turn T_2 OFF and T_1 ON.

For high speed circuits diode clamping can be included to prevent saturation, and also some improvement in switching

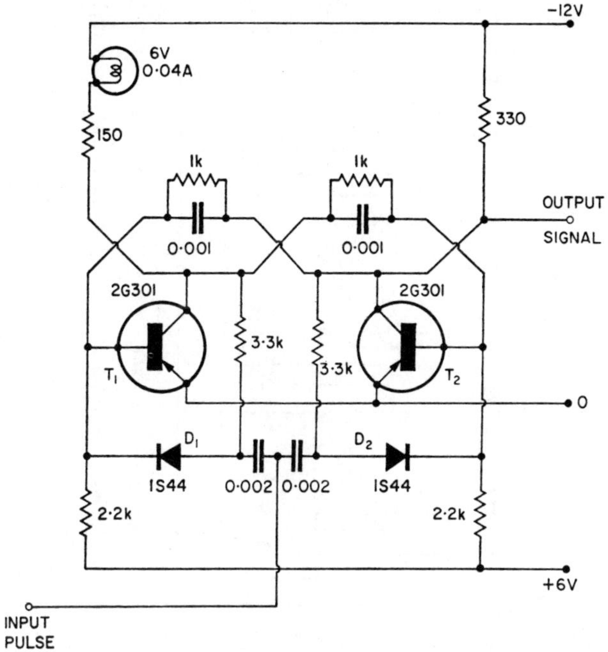

Fig. 13.10.—*Bistable circuit with pulse steering.*

time may be obtained by clamping with further diodes having opposite bias to prevent the transistors from going into cut-off.

COUNTING CIRCUITS

Binary counters

A simple basic counter circuit can be made by connecting several bistable multivibrator circuits in series. Taking the basic stage of Fig. 13.10, the output at the collector of T_2 will be a square wave with a frequency half that of the input

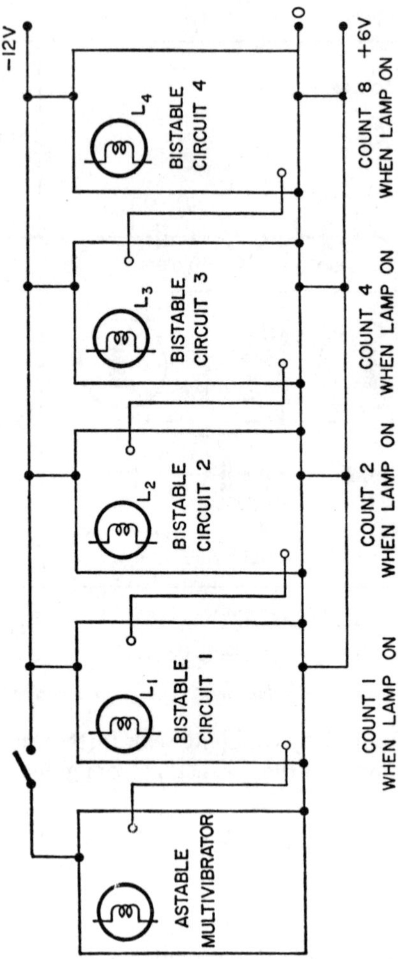

Fig. 13.11.—*Demonstration counter unit. Bistable circuits as Fig.* 13.10, *astable circuit as Fig.* 11.14.

pulses. This square wave can be differentiated and the positive pulse from the ON step used to trigger the next stage and so on.

To demonstrate this counting arrangement, four binary circuits can be arranged as shown in Fig. 13.11 and fed from an astable multivibrator of the type shown in Fig. 11.15 (often called a clock pulse generator in this application). When switch S_1 is closed, the astable multivibrator starts to switch on and off at about 1 c/s, giving a positive going output pulse each time its lamp lights. The first pulse will switch T_1 of the first bistable circuit OFF and T_2 ON. The second pulse switches T_2 OFF and T_1 ON, lighting lamp L_1 and giving a positive going output signal which is fed to bistable circuit 2 and switches its first transistor OFF, and second transistor ON and so on. The sequence will pass through the chain, the number of input pulses up to any moment being indicated by which lamps are ON, adding 1 when L_1 is ON, 2 when L_2 is ON, 4 when L_3 is ON and 8 when L_4 is ON. Thus this demonstration unit will count up to 16 and then start again.

The output information is more conveniently presented as a scale of 10 to give decimal information. To accomplish this, feedback is incorporated to feed extra pulses back into an earlier stage so that the four stages count 16 for only 10 input pulses. Then we have a scale-of-ten counter, and several such circuits can be connected in series to count up to any decimal number.

Ring counters

Another method of counting is to connect binary stages in a ring formation. In such a ring counter for decimal counting there are ten binary stages, one for each digit. Each stage feeds an indicator which shows when it is conducting, and with each input pulse the conducting stage is switched OFF and the next stage switched ON. The indicator system is arranged to give a direct "read out".

A convenient bistable circuit for use in ring counters is a pnp–npn combination as shown in Fig. 13.12. Assume T_1 is OFF. Then its collector will be at earth potential and the base of T_2 will be negative with respect to its emitter which is returned to a potential V slightly above earth. Thus T_2 will also be OFF. The collector of T_2 will be at a potential V_2 which is slightly more positive than V_1, and so T_1 will be held OFF. Thus the circuit will be stable in the OFF state. A negative trigger pulse at the input will drive transistor T_1 into conduction.

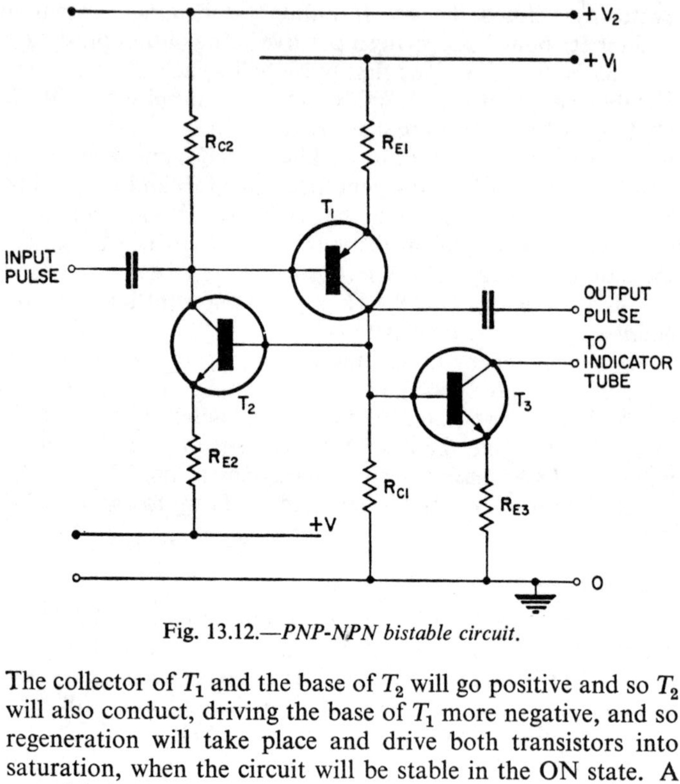

Fig. 13.12.—*PNP-NPN bistable circuit.*

The collector of T_1 and the base of T_2 will go positive and so T_2 will also conduct, driving the base of T_1 more negative, and so regeneration will take place and drive both transistors into saturation, when the circuit will be stable in the ON state. A negative input pulse will reverse the action and turn both transistors OFF.

The operation of a ring counter using this type of bistable stage can be described with reference to Fig. 13.13. Three stages of a ring are shown, the other seven stages being

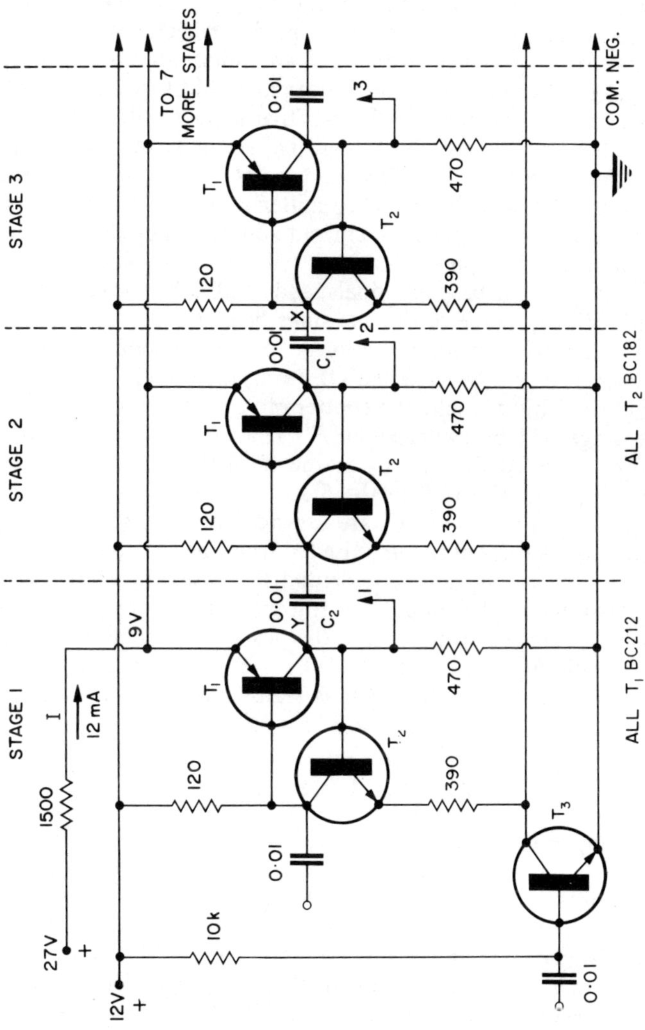

Fig. 13.13.—*Ring counting circuit.*

identical. The pnp emitters of each stage are fed from a common 12 mA constant current supply, and the emitters of the npn transistors are fed from the collector of a switch transistor (T_3) which feeds the switching impulses to the ring. The switch transistor is normally held ON by a 10 kΩ base resistor returned to the positive supply rail, and so its collector and the switch line will be nearly at zero potential. Assume that stage 2 is ON, and is taking the whole of the 12 mA constant current. The emitter of T_1 will be at 9 volts, determined by the constant current supply. Thus the emitters of all the other pnp transistors will be at 9 volts which is negative to their base potentials, and so all other stages will be held OFF by virtue of the one stage being ON.

If a negative impulse is applied to the input of the switch transistor it will be switched OFF and the switch line will go positive. This will take the emitter of T_2 (stage 2) positive, and the stage will be switched OFF. This will cause a negative pulse to be fed through C_1 to the third stage and a positive pulse through C_2 back to the first stage. The time constants associated with C_1 and C_2 are arranged to be long compared with the duration of the impulse fed from the switch transistor, so that when the impulse ends and the switch line goes back negative the negative potential still persisting at the point X is such that stage 3 switches ON, the positive potential persisting at Y holding stage 1 OFF. Thus with each impulse to the switching transistor the count is transferred one stage to the right.

COMPUTER LOGIC CIRCUITS

Transistor circuits are used extensively in computer circuits, but only passing reference can be made to them in this book. In performing logic operations two basic decision circuits are used. In each of these there is provision for several different input signals but only a single output signal. The first circuit gives an output signal only if all input signals are present simultaneously. This is called an AND gate since the condition for an output is that signal 1 AND signal 2 AND signal 3 etc. are all present together. The second circuit gives an output when any one of the input signals is present. This is called an OR gate since it gives an output when signal 1 OR signal 2 OR signal 3 etc. is present. In both of these circuits the polarity of the output signal is the same as the input signals.

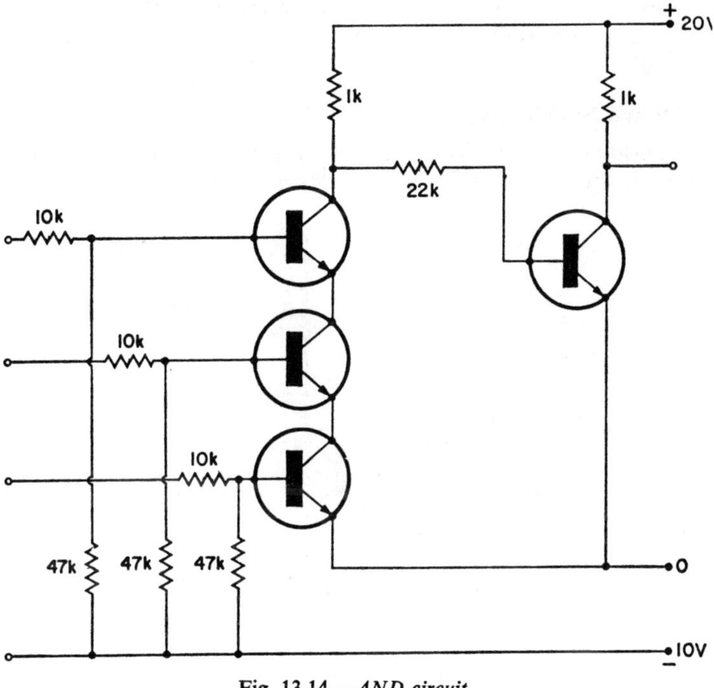

Fig. 13.14.—*AND circuit.*

To change the polarity of the output signal, a NOT circuit is used. By itself, this is a simple phase inverter circuit, such as a common emitter stage. When an AND function is combined with a NOT function the combination is called a NAND circuit, and when an OR function is combined with a NOT function we have a NOR circuit.

AND and NAND circuit

A simple AND circuit is shown in Fig. 13.14. The three transistors in series must simultaneously receive an input signal before the whole circuit can conduct and pass current through the resistance to give an output. By itself this series arrangement gives a NAND circuit since the polarity of the output pulse is opposite to that of the input. The addition of the common emitter stage converts the stage to an AND circuit.

OR and NOR circuit

Perhaps the simplest transistor logic circuit is the NOR circuit shown in Fig. 13.15. If a positive signal is fed to either

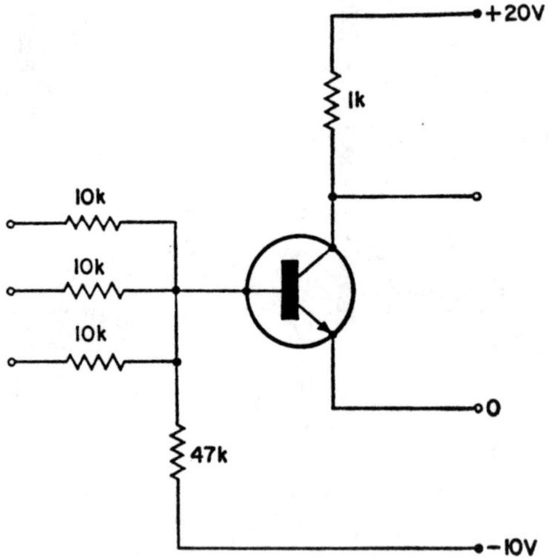

Fig. 13.15.—*Basic NOR circuit.*

input the transistor will conduct and give an output signal. The polarity of the output will be opposite to that of the input due to the common emitter arrangement. The circuit can be converted to an OR circuit by including a further common emitter stage to change the polarity of the output.

14
D.C. AMPLIFIERS

FOR many applications amplification of d.c. signals or very slowly varying signals is required. Amplifiers for this purpose are called *d.c. amplifiers*, and are generally divided into two types—"direct coupled", in which a d.c. connection is continued through the amplifier, and "chopper" types in which the d.c. signal is converted, by chopping, into a square wave a.c. signal, which is then amplified by a normal a.c. amplifier and converted back to d.c. by rectifying the output.

With a.c. amplifiers the minimum detectable signal is set by the noise generated at the input to the first stage. With direct coupled amplifiers a limit is reached before this, due to the drift of the d.c. parameters of the first transistor with change of temperature. Such changes will result in variation of the d.c. conditions in the transistor and so will be amplified along with the signal. The change is cumulative as the signal is amplified by stage after stage, as distinct from a.c. amplifiers in which any change in d.c. conditions is confined to each stage by the capacitive or inductive interstage coupling.

Thus in the design of d.c. amplifiers it is necessary to consider the magnitude of the transistor d.c. drift and to arrange that it is held to as low a level as required. For example if a signal of 0·1 volt is to be amplified with an accuracy of 1 per cent, the drift of the amplifier referred to the input must not exceed 1 mV.

Sources of drift

It is customary to specify drift in terms of the change of input d.c. voltage per degree centigrade necessary to maintain the output voltage constant with change of temperature. There are three major causes of drift in a transistor as the temperature is varied. First there is a change of emitter–base diode voltage for a fixed emitter current. With silicon transistors, this voltage decreases by approximately 2·5 mV per degree centigrade rise. Second is the change of collector leakage current which doubles for about every 11 degrees centigrade rise in temperature. Third is the change of current gain h_{fe}, which increases by about 1 per cent per degree centigrade.

It is also possible that drift can result from long term ageing effects in the transistor, but modern manufacturing methods such as the planar process have generally reduced such effects to a negligible level.

The most important of the sources of drift is undoubtedly the base–emitter voltage V_{BE}. However while it is the largest of the three it is quite stable and is consistent between transistors of a given type, and so it is possible to design circuits to include some form of balancing compensation.

With regard to the drift due to collector leakage current, the very low value of leakage current with silicon transistors reduces the effect to a negligible level. For this reason it is now generally recommended that only silicon transistors be considered for d.c. amplifier work and, even further, silicon planar transistors offer the best performance and long term reliability.

As to drift of current gain the effect of this can be minimised by including sufficient negative feedback so that the error introduced is reduced to a level lower than that due to the base–emitter voltage drift. Negative feedback will, in any case, usually be required with d.c. amplifiers to allow for the difference in gain between individual transistors, so that the required d.c. operating currents and gain are maintained.

Thus, assuming the use of silicon planar transistors and negative feedback, the main problem associated with drift is that of compensating for the change of base–emitter voltage V_{BE} with temperature. The degree of compensation required will, of course, depend upon the magnitude of the smallest signal to be amplified.

DIRECT COUPLED AMPLIFIERS

Interstage coupling

In addition to the drift problem there is one other major consideration in direct coupled amplifiers, the method of connecting the output of one stage to the input of the next. This coupling not only passes on the signal but must also be designed so that the d.c. potential at the output of the first transistor is correct for establishing the bias conditions required by the second transistor. The potential at the collector of the first stage will usually be higher than that required at the base of the next stage, and special means must be provided so that the signal can be coupled without substantial loss.

Direct coupled pair (Darlington pair)

A very convenient method of arranging a two stage direct coupled amplifier is shown in Fig. 14.1. This arrangement is known by a variety of names—Darlington pair (after

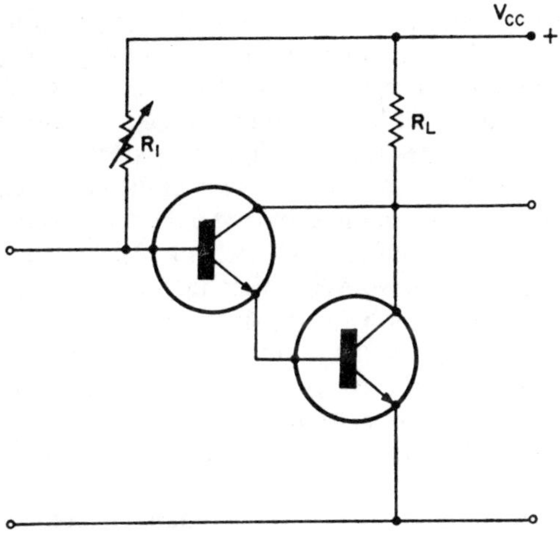

Fig. 14.1.—*Darlington direct coupled circuit.*

S. Darlington who proposed it in 1953), compound connected pair, double emitter follower, super alpha transistor and so on. Both transistors are operated in the common emitter mode, signal current is fed to the base of the first transistor, and the resulting amplified emitter current is fed direct into the base of the second transistor. The overall current gain is equal to the product of the common emitter current gains (at the operating currents) and the input impedance is very high—higher than that of a single transistor by a factor roughly equal to the current gain of the first transistor.

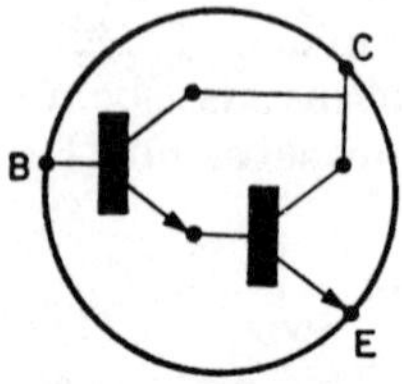

Fig. 14.2.—*Composite direct coupled transistor.*

The combination is very convenient for low noise input circuits. For such use the first stage must be operated at a very low collector current. The circuit is also very useful at higher current levels for the output stages of an amplifier, providing a very economic arrangement for the driver and output stages.

The d.c. bias is set by adjusting the base current to the first transistor. As an example, using transistors with operating current gains of the order of 50, the overall current gain will be about 2,500. If the required output collector current is 5 mA, the input base bias current will be only 2 μA. Thus the base bias circuit must have very high impedance and this may impose some limitation in use since with a d.c. amplifier the input circuit must be arranged to include the bias arrangements. It must also be remembered that the first transistor will be operating with a low collector current (100 μA in the above case) and so its current gain will be lower than for normal operation.

The circuit can be used in the common collector mode with a load resistance in the emitter circuit of the second transistor to give a very high value of input resistance. A limit is imposed, however, by the shunting effect of the collector resistance of the first transistor. Input resistances of the order of one megohm can be obtained.

The arrangement can be treated as a single device as shown in Fig. 14.2—in fact some manufacturers supply units consisting of two transistor wafers assembled into a single unit (e.g. 2N997). Advantage can be taken of the fact that by using two very ordinary low gain transistor wafers a useful high gain unit can be made.

Direct collector to base coupling

Various methods of coupling the output from the collector of one stage to the base of the next stage can be used. In the first method, shown in Fig. 14.3, a small resistance R_E is

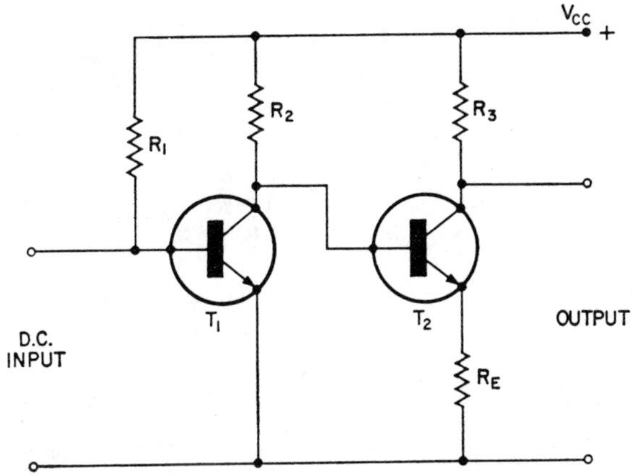

Fig. 14.3.—*Two-stage direct coupled amplifier.*

included in the emitter circuit of the second stage to allow the base to operate at the same potential as the collector of the first stage. This arrangement is not ideal, since the resistor will introduce signal loss due to negative feedback, and if it is made small to minimise this loss the collector voltage of the first stage must be correspondingly low. A better arrangement is to include a zener diode in the second emitter as shown in Fig. 14.4. The collector voltage of the first transistor T_1 will be equal to the voltage across the zener diode plus the base–emitter voltage of transistor T_2 (approximately 1 volt for silicon transistors). Thus the zener diode should be chosen to have a voltage 1 volt lower than the collector voltage required for transistor T_1. Resistance R_4 ensures that the current through the zener diode is sufficient to bias it into its low resistance breakdown region; then the impedance to the signal is small and any negative feedback is correspondingly small. If the emitter current of transistor T_2 is sufficient to give this condition, resistor R_4 can be omitted.

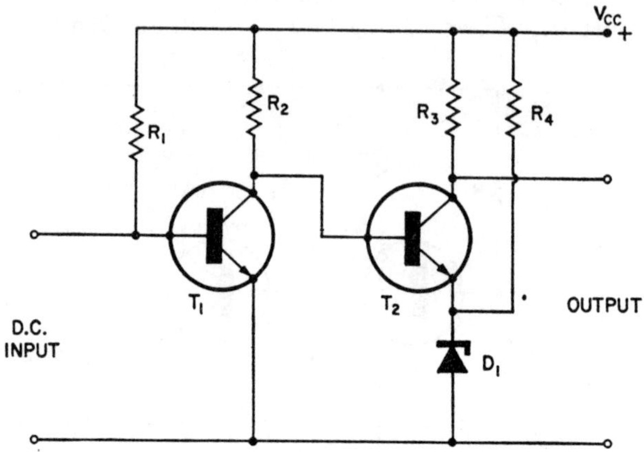

Fig. 14.4.—*Two-stage d.c. amplifier with zener diode bias.*

A practical example of an amplifier using zener diode biasing is shown in Fig. 14.5. This shows a three-stage amplifier with overall negative feedback. The amplifier has an open loop gain of about 2,500, the actual operating gain being set by adjusting the amount of negative feedback. The second-stage emitter voltage is set by a 4·7 volts zener diode, allowing the collector of the first stage to operate at 5·7 volts. With a collector resistance of 22,000 ohms the first transistor will have a collector current of 0·75 mA. The emitter voltage of the third transistor is set by a 6·8 volts zener diode giving a collector voltage for the second transistor of 7·8 volts and a collector current through the 6,800 ohms resistor of 2·2 mA. A further 2·6 mA of current is fed into the zener diode to give a total diode current of 4·8 mA.

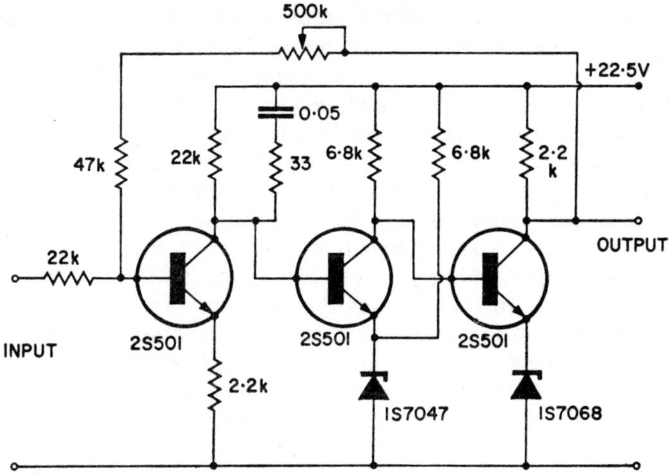

Fig. 14.5.—*Three-stage d.c. amplifier.*

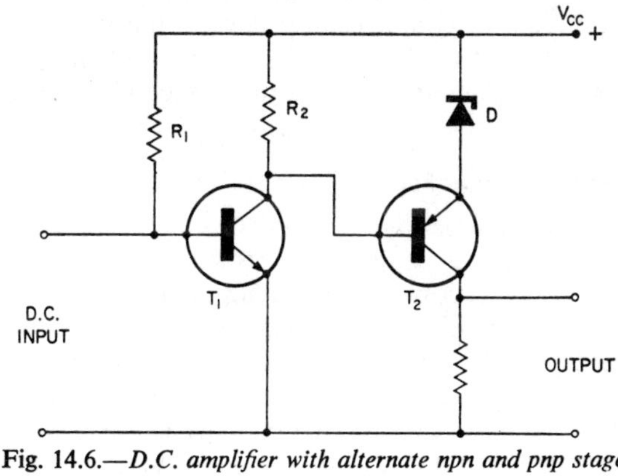

Fig. 14.6.—*D.C. amplifier with alternate npn and pnp stages.*

Another arrangement that can be used to advantage has alternate npn and pnp stages as shown in Fig. 14.6. The collector of the first transistor T_1, an npn transistor, is connected direct to the base of T_2, a pnp transistor. Control of the collector current of transitor T_2 can be achieved by varying R_1 and R_2.

A particularly interesting feature of this arrangement is that it avoids the amplification of residual a.c. ripple on the collector supply. With the arrangement of Fig. 14.3 ripple voltage on the collector supply will result in a current flowing down R_2 and through the base emitter junction of T_2, and so an amplified ripple will appear at the collector of T_2. With the npn–pnp combination such ripple current cannot flow in the base–emitter circuit of T_2 and so it is not amplified.

In the various arrangements described above, no attention has been paid to minimising or compensating for the drift with temperature. In such circuits the magnitude of the drift will mainly depend upon the value of source resistance. If this is high, the effect of any change of base–emitter voltage will be swamped as we effectively have constant current conditions and so the drift due to the first transistor will be reduced to a low value. If the source resistance is low, the full base–emitter voltage drift of 25 mV per degree centigrade will be effective. To reduce the drift with low source resistance, it is necessary to use a compensating circuit, the most common arrangement being a difference (or differential) amplifier, in which a second transistor is used to balance out changes due to drift.

Long-tailed pair amplifier

The basic compensating circuit, called a long-tailed pair, or sometimes an emitter coupled pair, is shown in Fig. 14.7. To consider the operation of the circuit assume that the two transistors are identical in all respects and that the circuit components are exactly symmetrical. Emitter resistance R_3 is made large and the potentials of the base electrodes of the transistors are fixed by the base voltage dividers. Thus the potential across R_3 must also be fixed, and the sum of the two emitter currents, flowing through R_3, must remain constant. Now, if equal *in phase* or *common mode* signals are applied to the two base electrodes no output will be produced since the total emitter current cannot change. If, however, a "push–pull" input is applied across the two base terminals the current through one transistor will increase and that through the other will decrease, the total emitter current remaining constant. Then an output signal will appear between the two collectors. Referring again to Fig. 14.7, with no signal input the potentials at the two collectors will be identical, and so there will be no

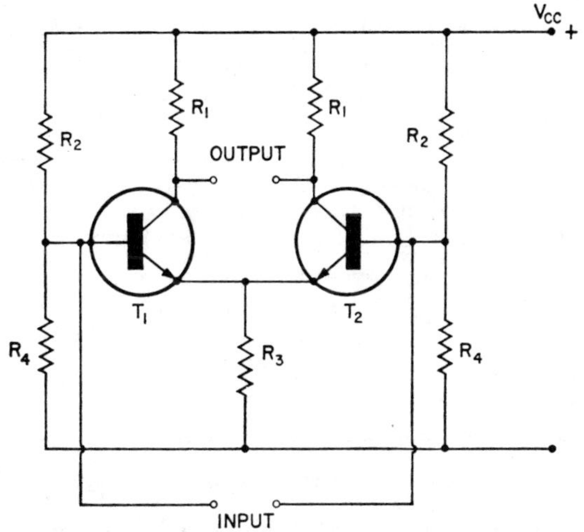

Fig. 14.7.—*Long-tailed pair differential amplifier.*

output. If the temperature increases, the base–emitter voltages of the two transistors will decrease by similar amounts, giving a common mode change and there will be no output. Thus as long as the two transistors have equal changes of V_{BE} with temperature the drift will be balanced out. When a d.c. signal is applied it will divide equally between the two base–emitter junctions, so that their bias points are moved in opposite directions. The collector current of T_1 will increase and that of T_2 will decrease, resulting in a signal output between the two collectors.

In practice transistors will differ slightly in base–emitter characteristics, and so there will be some resultant drift, but it will be considerably lower than with unbalanced circuits. With transistors of a given type, that is made by the same process such as the diffused planar method, the change of base-emitter voltage is very consistent from one transistor to another. Thus in a long-tailed pair circuit the two transistors should be of the same type and care should be taken to ensure

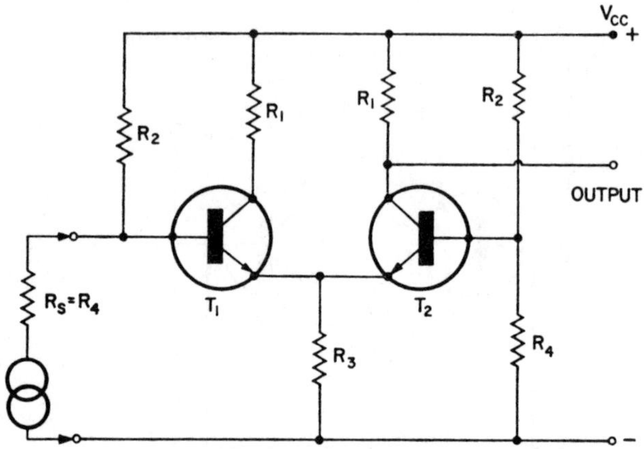

Fig. 14.8.—*Single-ended long-tailed pair stage.*

that the temperatures of the two transistors are as equal as possible. This can be carried out by mounting the two transistors side by side on a common heat sink.

More recently dual transistors have become available such as the 2N2639 type. This consists of two transistor wafers mounted side by side in a single encapsulation, the wafers normally being taken from the same diffused slice. In another version the two transistor elements are formed in a single wafer of silicon. With each type the characteristics and temperature variations of the two transistors will be nearly identical.

Push–pull signals are not always convenient, and with some slight compromise in stage gain and drift performance a single ended version of the circuit can be used as shown in Fig. 14.8. When the input signal is applied it will divide between the two base–emitter junctions and resistance R_4. The voltages across the two base–emitter junctions will bias the

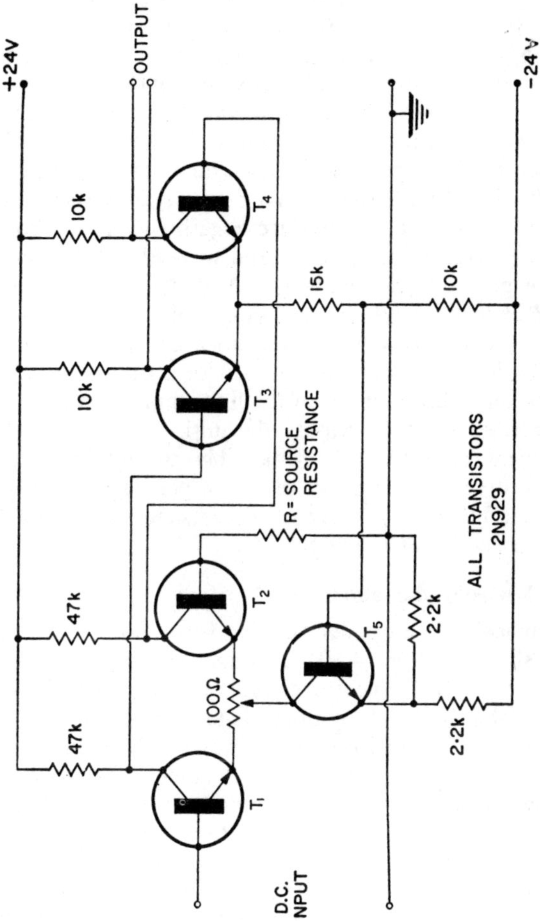

Fig. 14.9.—*Two-stage long-tailed pair differential amplifier.*

transistors in opposite directions, one current will increase and the other will decrease. Thus an output signal can be taken from the collector of transistor T_2. A typical drift performance of this single ended type circuit is 100 μV per degree centigrade.

Increased gain can be obtained by adding further long-tailed pair stages, and the drift can be reduced by what is known as *common mode feedback*. The amount of the residual drift due to temperature change depends upon how constant the voltage across R_3 (Fig. 14.8) remains. To improve the stability of this voltage it is possible to introduce negative feedback from the common emitter point of the second stage (T_3, T_4, Fig. 14.9) to the same point in the first stage. Any change in the voltage across R_3 will be amplified and will result in a larger voltage across the common emitter resistance of the second stage. This variation is fed back to the base of a transistor (T_5) connected in the emitter circuit of the first stage. The feedback is in the correct sense to oppose the initial change. With this arrangement (Fig. 14.9) it is possible to obtain drift performance down to the order of 20 μV per degree centigrade. Output can be taken across the two collectors of T_3 and T_4, or from either collector to earth.

Typical low noise d.c. amplifier

A practical d.c. amplifier suitable for a variety of applications is shown in Fig. 14.10. This amplifier illustrates a combination of several types of circuit and has been mentioned earlier in the book. The input stage uses a 2N2500 silicon planar field effect transistor which gives a high input resistance, about 10 MΩ, with low noise performance. The second stage is a long-tailed pair of 2N929 npn silicon planar transistors, the output of which is connected to a pnp stage (2N2411) followed by an npn amplifier stage and an npn emitter follower output stage, the last two stages using more 2N929 silicon planar transistors. The overall voltage gain of the amplifier is 100, and the input noise figure with a source resistance between 100 kΩ and 1·5 MΩ is as low as 1 db. The frequency reponse with a 1·5 MΩ source resistance is d.c. to 10 kc/s and the output resistance 500 ohms.

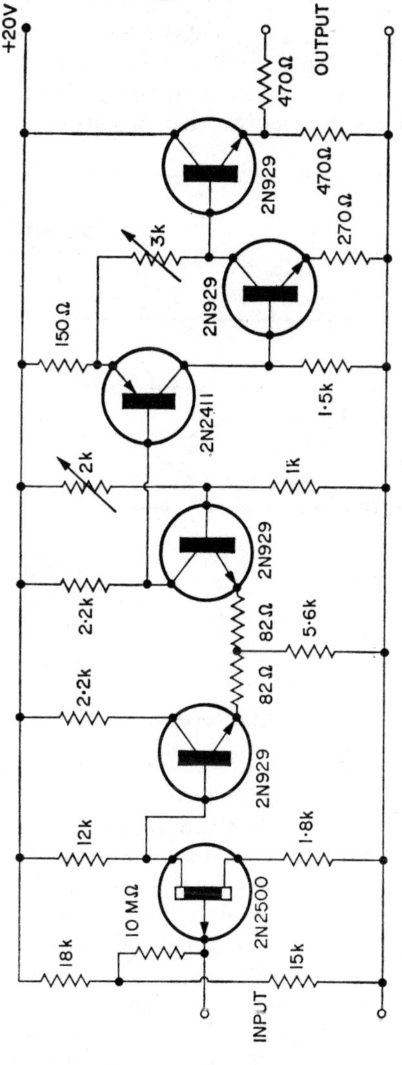

Fig. 14.10.—*Low noise d.c. amplifier.*

CHOPPER CIRCUITS

As stated earlier an a.c. coupled amplifier is not subject to cumulative drift output, and to take advantage of this an alternative method of amplifying d.c. signals is to chop the d.c. signal into a square-wave a.c. signal which is amplified

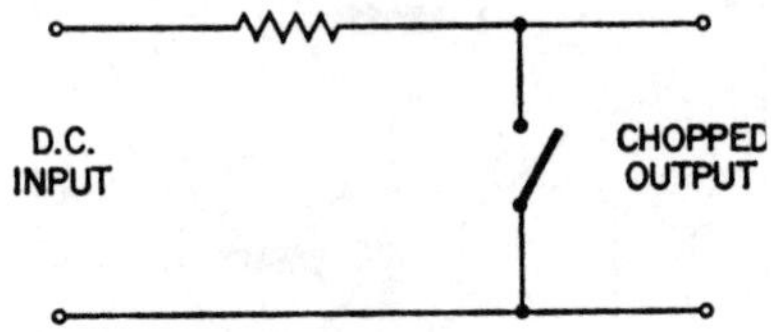

(a) Shunt chopper

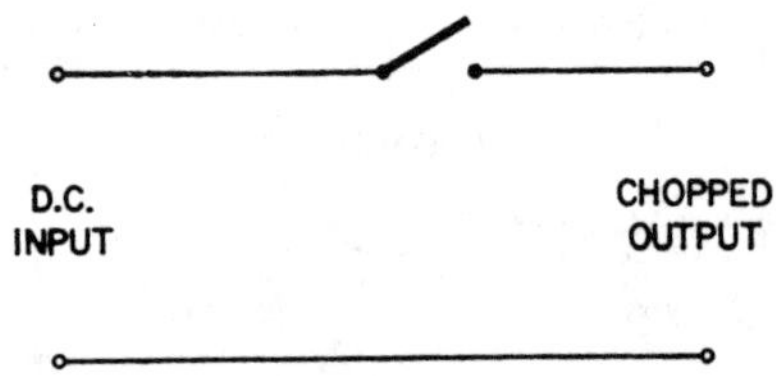

(b) Series chopper

Fig. 14.11.—*Basic chopper circuits.*

by an a.c. amplifier and finally rectified back to d.c. To convert the d.c. signal to a.c., a fast acting switch is included in the input circuit to alternately pass and block the signal at a convenient frequency. This produces a square-wave signal with peak amplitude proportional to the d.c. input signal. The switching process is called *chopping* and the overall equipment is called a *chopper amplifier*.

The simplest form of chopper switch is a vibrating unit, such as a high speed carpenter relay. There are two basic switching arrangements, as shown in Fig. 14.11. The first, with shunt contacts, short-circuits the input signal to reduce it to zero, and the second, with series contacts, open-circuits the signal. Such mechanical systems, however, have the obvious disadvantages of contact troubles and switching frequency

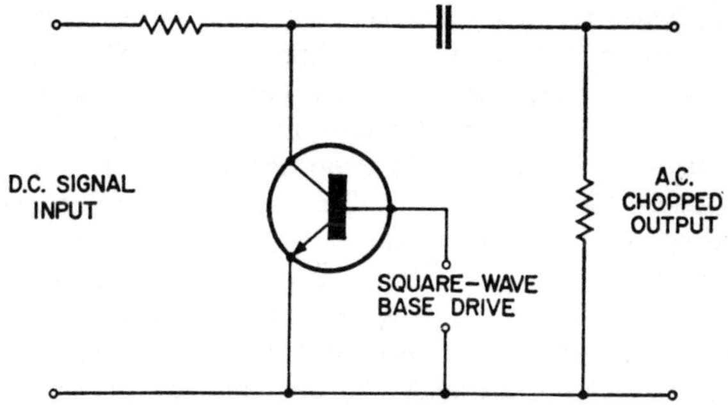

Fig. 14.12.—*Transistor shunt chopper.*

limitations. These disadvantages can be avoided by the use of a transistor, which can be operated as a very effective switch, as described in Chapter 13.

When a transistor is operated as a switch, the collector and emitter become the switch terminals and switching is accomplished by applying a suitable voltage between the base and emitter. A basic transistor chopper circuit using shunt switching is shown in Fig. 14.12. The d.c. signal is fed through a series resistance and the chopper transistor connected in shunt as shown. Thus the signal becomes the collector supply voltage. A square-wave driving signal is applied to the base electrode so the the transistor is either cut off or saturated. When the transistor is cut off, its resistance is very high and so its shunting effect is small. When the transistor is saturated, its resistance is very low and the voltage across collector to emitter falls to a low value. Thus the signal at the output terminals will be a square wave with amplitude approximately equal to the value of the d.c. input.

Although the collector leakage current and the collector saturation voltage are both small they are a measure of the departure from the ideal and so present a limit to the performance of the system. In the OFF condition the collector leakage current will flow through the series resistance and so will introduce a small loss of signal. In the ON condition the voltage across the transistor will fall to a value just larger than a value called the *offset* voltage ($V_{CE(\mathrm{offset})}$) which is effectively the saturation voltage at zero emitter current. This voltage is produced by the base drive current flowing through the transistor. (At high emitter currents the saturation voltage increases due to the voltage drop across the series collector resistance.) The value of the offset voltage can be reduced by operating the transistor in the inverse direction, that is with the collector and emitter reversed. In addition to the offset voltage (now $V_{EC(\mathrm{offset})}$) being reduced, the current gain in the inverse direction is much lower and so an increased base drive is necessary. With a typical silicon planar chopper transistor such as the 2N2432, the inverse offset voltage is about 0·5 mV and the collector leakage current of the order of 1 nanoampere.

Balanced chopper circuit

Various circuits have been developed to compensate for the effects of leakage current and offset voltage. A typical circuit is shown in Fig. 14.13. Two transistors are used, and it is arranged that they are switched on and off together. The offset voltages of the two transistors in the ON state appear in series opposition, and an exact balance can be obtained by adjustment of potentiometer P_2. Similarly the leakage currents on the OFF state are also in opposition, and adjustment of potentiometer P_1 will give the OFF balance.

Double emitter chopper

Modern transistor techniques have resulted in a special single transistor which will do the same work as the two units in Fig. 14.13 more efficiently. This is the double emitter chopper (Type 3N74), a device with a single collector, a single base and two symmetrical emitters. Its use in a chopper circuit is shown in Fig. 14.14, the device being operated in the inverted arrangement with the two emitters acting as collectors. Since

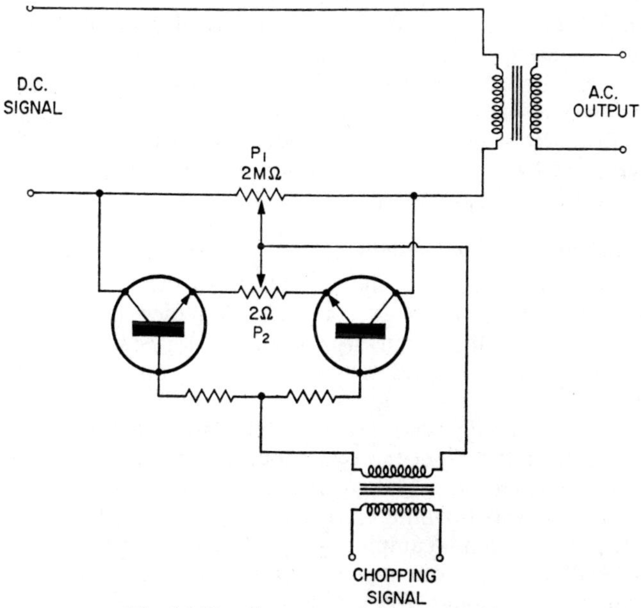

Fig. 14.13.—*Compensated chopper circuit.*

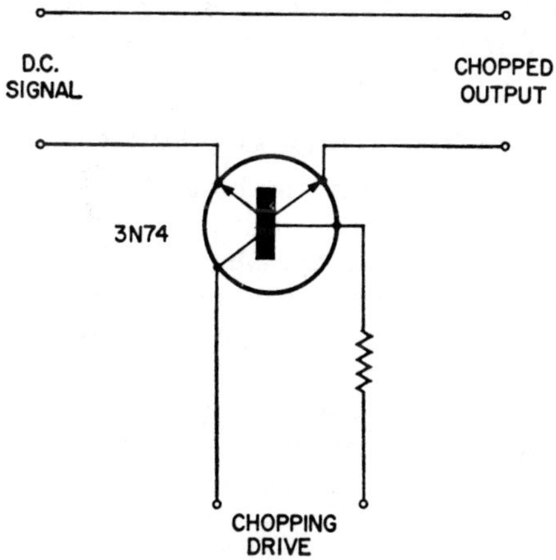

Fig. 14.14.—*Dual emitter chopper transistor.*

the emitters are symmetrical and part of a single structure, the corresponding offset voltages and their thermal variation will be very similar. In practice, the type 3N74 transistor has a residual offset voltage—the difference between the two individual voltages—of 50 μV. The transistor is a silicon planar type and its leakage current, of the order of a nanoampere, is small enough to be ignored.

A chopper stabilised amplifier

A typical example of the application of the double emitter transistor in a chopper stabilised amplifier is shown in Fig. 14.15. A 3N74 transistor is used with a shunt switching arrangement that gives two identical a.c. outputs for zero input signal. A d.c. input signal will result in an unbalance, which is amplified by an a.c. differential amplifier. Any identical output from the two chopper outputs due to offset voltage changes or switching spikes will appear as a common mode signal to the differential amplifier and will be rejected. After the differential amplifier, a single transistor demodulator is used to revert the signal back to d.c. The transistor is switched in antiphase to the chopper transistor. The switching drive can conveniently be carried out by using a square wave multivibrator, taking the output from one collector for the chopper and from the other collector for the demodulator drive.

Use of the field effect transistor as a chopper

The field effect transistor can be used as an efficient shunt chopper and in some aspects can offer advantage. Since the ON condition is produced by reducing the reverse voltage on the gate electrode, rather than increasing the base current as in the normal transistor, there is no offset voltage. Thus the ON voltage is smaller and more stable. Also the field effect transistor can result in lower noise operation as the current flow is by majority carriers and no recombination is involved.

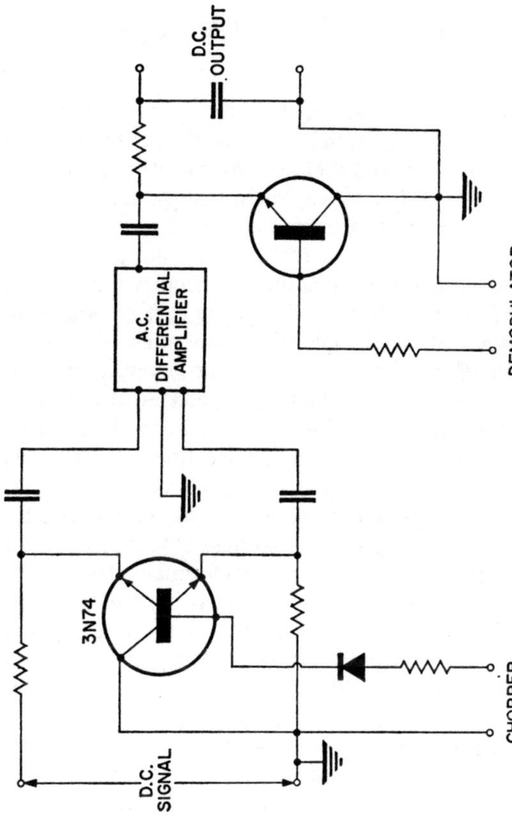

Fig. 14.15.—*Chopper stabilised amplifier.*

15

POWER SUPPLIES

In addition to the basic requirement of providing a suitable supply for operating transistor circuits, consideration of the design of power supplies includes several other aspects, such as the use of transistors and diodes in voltage stabilising circuits, in d.c. to a.c. inverters and in d.c. to d.c. converters. Primary batteries are often used as the collector supply for transistor circuits, and in many applications, particularly portable equipment and low level circuits, this is the best arrangement from several points of view, especially economics. In the majority of commercial transistor equipment, however, mains operation will be required. The production of rectified supplies from the a.c. mains in general follows the established practice, using step-down transformers and silicon rectifiers. Since transistors usually operate with relatively low collector voltage, medium and high power operation demands quite high current swings, and so the subject of output voltage regulation must receive special attention. Considerable developments have taken place in the design of voltage stabilising circuits, and it is significant that these stabilising circuits form an important application of transistors, not only for stabilised supplies for transistor circuits but also to provide constant and reference voltages for many other items of electronic equipment.

Transistor circuits offer considerable advantage over the previously used mechanical vibrators in d.c. to d.c. converter units, and high power transistors can be effectively used in place of rotating machines to give low and medium power a.c. outputs from a d.c. source.

These various aspects of power supplies will all be covered in this chapter, with emphasis on the use of transistors in stabilising circuits, d.c. to a.c. inverters and d.c. to d.c. converters.

Rectifier circuits

There are various standard rectifier circuits, the choice depending upon the required d.c. voltage and load current, and on economic considerations coupled with the required reliability. Since the voltage required for transistor circuits is normally low, in the range of 10–100 volts, a step-down mains transformer will generally be used.

Half-wave rectifier circuit

This is the simplest rectifier circuit, and is shown in Fig. 15.1 (a). If the r.m.s. voltage of the transformer secondary is E, the d.c. voltage across the reservoir capacitor will be $\sqrt{(2)}\,E$ with no load, falling to approximately E at full load, depending upon the value of the capacitor. During the non-conducting half-cycle, the peak reverse voltage appearing across the rectifier will be the sum of the peak values of the a.c. input voltage and the voltage across the capacitor. This can have a maximum value of $2\sqrt{(2)}\,E$. The a.c. ripple appearing on the d.c. output will have a frequency equal to the mains frequency.

Full-wave rectifier circuit

The full-wave circuit is illustrated in Fig. 15.1 (b). A centre-tapped secondary winding is used, with two rectifiers connected so that they each use one half of the a.c. input. One rectifier delivers current to the load during the "odd" half-cycles, and the other during the "even" half-cycles. If E is the r.m.s. a.c. voltage across each half of the secondary winding, the d.c. output voltage across the capacitor will again be $\sqrt{(2)}\,E$ with no load, but will only fall to the order of $1{\cdot}2E$ on load, since the second rectifier helps to hold the voltage up. As there are two rectifiers supplying current to the load, each rectifier will only be passing one half of that passed by the single rectifier in the half-wave case. The ripple on the d.c. output voltage will be twice that of the mains frequency.

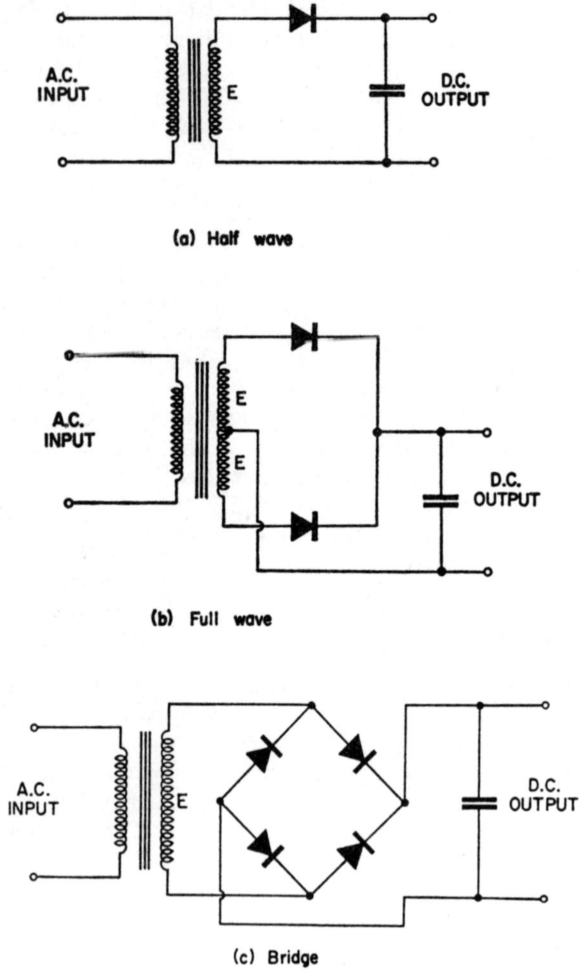

Fig. 15.1.—*Basic rectifier circuits.*

Bridge rectifier

The bridge circuit is shown in Fig. 15.1 (c). The use of this circuit gives full-wave rectification without a centre tapped winding. The peak reverse voltage appearing across the rectifier cells is equal to the peak value of the a.c. input voltage, i.e. $\sqrt{(2)}\,E$. This is only one-half of the value with the previous full-wave circuit, and so for a given value of output voltage the rectifier cells need have only half the voltage rating. To offset this, four rectifier cells are needed instead of two.

Voltage doubler circuit

In the full-wave voltage doubler circuit Fig. 15.1 (d), two rectifiers are connected so that the d.c. output is twice that of the other circuits. During the "odd" half-cycles C_1 is charged

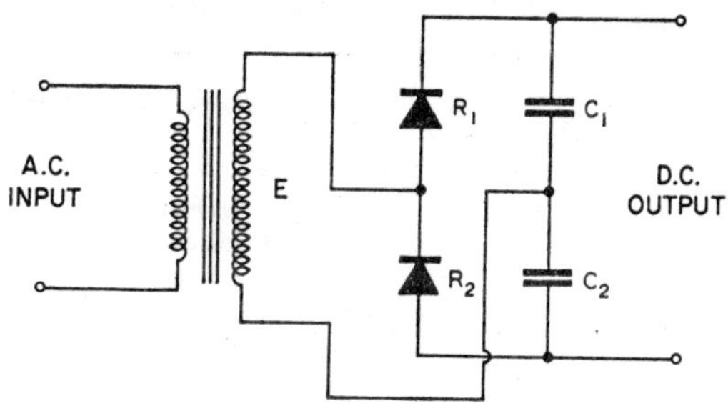

(d) Voltage doubler

Fig. 15.1 (continued).—*Rectifier circuits.*

to a voltage $\sqrt{(2)}\,E$ (no load) through rectifier R_1 and during the even half-cycles C_2 is charged to a similar voltage through R_2. Thus the total d.c. output voltage is $2\sqrt{(2)}\,E$ falling to about $2E$ on load. The peak reverse voltage across each rectifier is $2\sqrt{(2)}\,E$.

The various characteristics of the above rectifier circuits in terms of the input a.c. voltage E are summarised in Table 15.1 and in Table 15.2 some actual values are given for a d.c. output requirement of 24 volts, 0·5 amp.

Table 15.1. Rectifier circuit characteristics

Circuit	*Input voltage* (r.m.s.)	*Output voltage no load*	*Approx. output voltage full load*	*Rectifier peak reverse voltage*	*Rectifier current*
Half wave	E	$\sqrt{(2)}\,E$	E	$2\sqrt{(2)}\,E$	I
Full wave	$E+E$	$\sqrt{(2)}\,E$	$1{\cdot}2E$	$2\sqrt{(2)}\,E$	$0{\cdot}5I$
Bridge	E	$\sqrt{(2)}\,E$	$1{\cdot}2E$	$\sqrt{(2)}\,E$	$0{\cdot}5I$
Voltage doubler	E	$2\sqrt{(2)}\,E$	$2E$	$2\sqrt{(2)}\,E$	I

Table 15.2. Rectifier circuit values

Circuit	*D.C. output voltage* (0·5 A)	*Required input voltage* (r.m.s.)	*Required rectifier rating*		*Number of rectifiers*
			Reverse voltage	*Current*	
Half wave	24 V	24 V	70 V	0·5 A	1
Full wave	24 V	20+20 V	60 V	0·25 A	2
Bridge	24 V	21 V	30 V	0·25 A	4
Voltage doubler	24 V	12 V	35 V	0·5 A	2

Smoothing circuits

The amplitude of the ripple voltage appearing on the d.c. output is reduced by including a low pass filter consisting of an inductance–capacitance, or resistance–capacitance combination, as shown in Fig. 15.2. The expressions for the ripple reduction ratio are given. The resistance smoothing element will, of course, introduce a higher drop in voltage than the inductance, and will result in more severe regulation, and so normally is only used for low current, constant load applications.

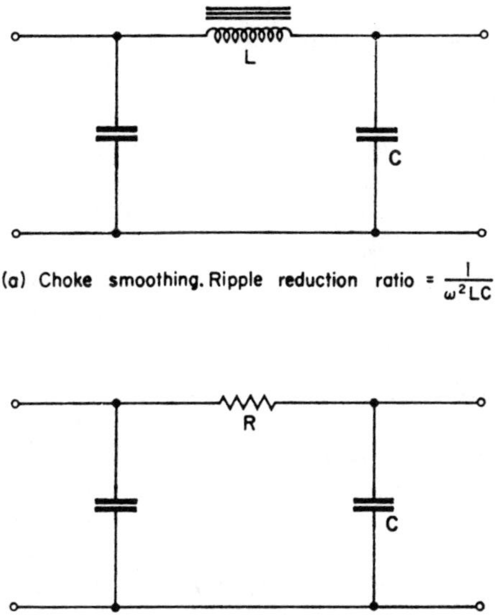

(a) Choke smoothing. Ripple reduction ratio = $\frac{1}{\omega^2 LC}$

(b) Resistance smoothing. Ripple reduction ratio = $\frac{1}{\omega CR}$

Fig. 15.2.—*Smoothing circuits.*

Voltage stabilisation

As the load current taken from a normal rectifier circuit is increased the output voltage will fall, partly due to the ohmic resistances of the transformer windings, the rectifier cells and the smoothing elements, and partly due to the lower effective storage efficiency of the reservoir capacitor with the lower load resistance. In many applications the load current will vary, as, for example, in class B and switching circuits, and it is then desirable to include some form of stabilisation to maintain the output voltage constant. Various arrangements can be used depending upon the degree of stabilisation required.

Zener diode shunt stabiliser

The zener diode stabiliser has the advantage of simplicity, and for many applications, particularly for low current loads, it can give sufficient stabilisation. The arrangement is shown in Fig. 15.3 (a). The zener diode is connected in the reverse

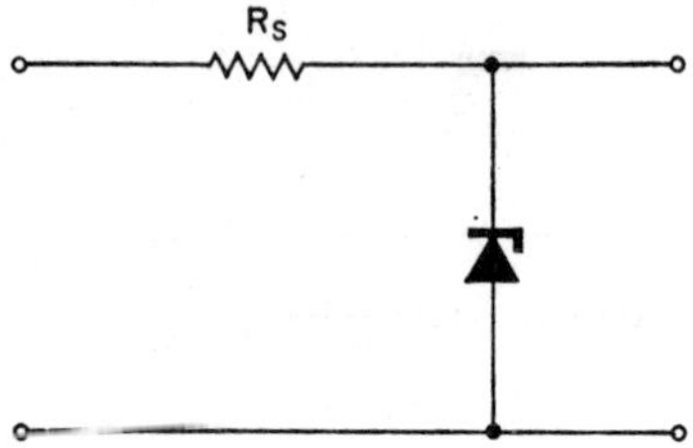

(a) Zener diode stabiliser

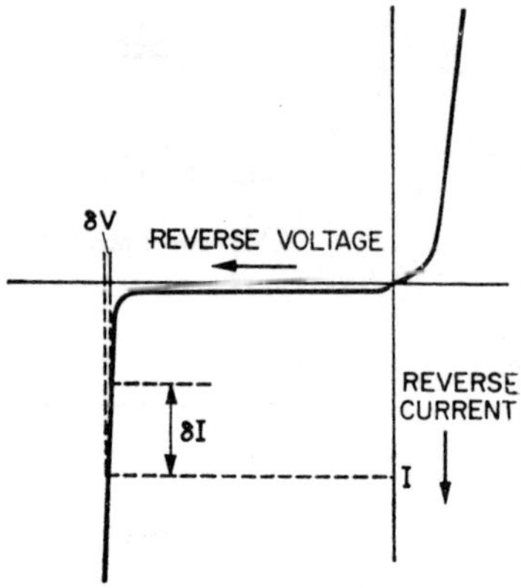

(b) Zener diode characteristics

Fig. 15.3.—*Zener diode stabilisation.*

direction across the load and is fed from the unstabilised voltage through a series resistance R_S. The characteristic of a zener diode is illustrated in Fig. 15.3 (b). In the reverse direction the current is small until the breakdown is reached, and then the current increases very rapidly for a small increase of voltage. The diode is chosen to have a breakdown voltage equal to the required output voltage, and the input voltage and series resistance are set so that the diode is biased in the breakdown region at a predetermined current such as the value I in Fig. 15.3 (b). Now if the load current increases the output voltage will tend to fall, but this will result in a decrease in the diode current such that the sum of the load plus diode currents remains almost constant. Thus the output voltage will remain almost constant, the only voltage change being δV due to the change in the diode current δI.

Similarly if the input supply voltage increases the voltage across the zener diode will tend to increase, causing an increased current through the zener diode. This extra current will produce a voltage drop across the series resistance R_S almost equal to the increase of input voltage, thus keeping the output voltage almost constant.

At maximum load current and minimum input voltage there must still be sufficient current flowing through the zener diode to hold it in the breakdown region, at least 2 or 3 mA. If this value is I_M and the maximum load current I_L, the total current flowing through the series resistance R_S will be (I_M+I_L). If

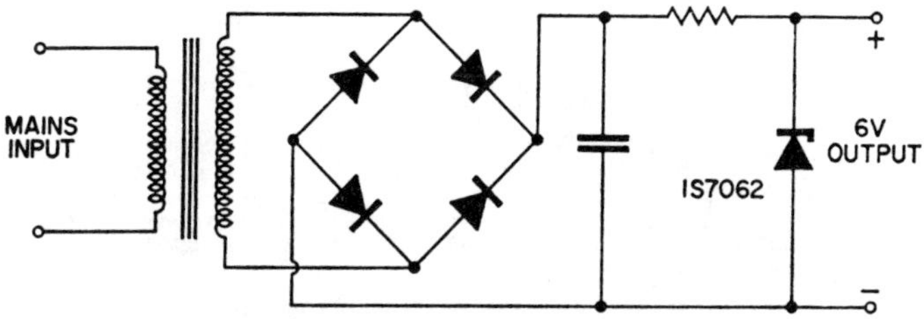

Fig. 15.4.—*Zener diode shunt stabiliser.*

the minimum input voltage is V_M and the stabilised output voltage V_O the value of R_S must be equal to $(V_M-V_O)/(I_M+I_L)$. For good stabilisation the value of V_M should preferably be at least 1·5 times the value of V_O.

A practical example of a zener diode shunt stabiliser circuit is shown in Fig. 15.4. This circuit gives an output voltage of 6 volts. It allows an input voltage variation of ± 15 per cent and a load current variation of 5 to 50 mA with a maximum output voltage change of only 0·1 volt.

For higher output voltages, diodes with correspondingly high breakdown voltage can be used, but the value of the diode slope resistance in the breakdown region increases faster than the value of breakdown voltage, and better performance can be obtained by connecting two or more low voltage (5–6 volts) zener diodes in series to obtain the required output voltage. This series arrangement also gives the optimum performance from the viewpoint of stability of the breakdown voltage with temperature changes as the change is nearly zero with diodes in the 5–6 volts region. The choice of method is entirely dependent upon the degree of stabilisation required and of course upon economic considerations.

Transistor shunt stabiliser

An improvement over the diode shunt stabiliser can be obtained with a transistor circuit as shown in Fig. 15.5. A transistor is connected across the supply, with its base electrode fed through a zener diode. The input voltage and base resistance are again set so that the zener diode is biased into

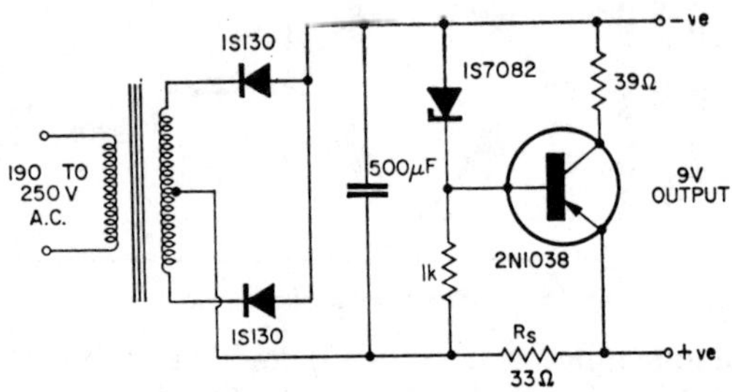

Fig. 15.5.—*Transistor shunt stabiliser.*

the breakdown region. If the input voltage increases, the zener diode passes more current, which increases the base current of the transistor and causes a much larger current to flow down through the transistor. This increase of transistor current results in an increased voltage drop across the series resistance R_S so that the output voltage is maintained almost at its original value.

The circuit shown in Fig. 15.5 was designed to give a stable 9 volts supply from the a.c. mains to replace the battery in portable transistor radio receivers. A mains voltage variation from 190 to 250 volts will result in an output voltage change of only 0·2 volt and a load current variation from zero up to 125 mA gives an output voltage change of 0·25 volt.

Series voltage stabilisers

In series stabilising circuits it is arranged that the voltage change of an unstabilised supply appears across a series control transistor so that the output from the transistor remains constant. The simplest arrangement of this type of stabiliser is the emitter follower circuit as shown in Fig. 15.6.

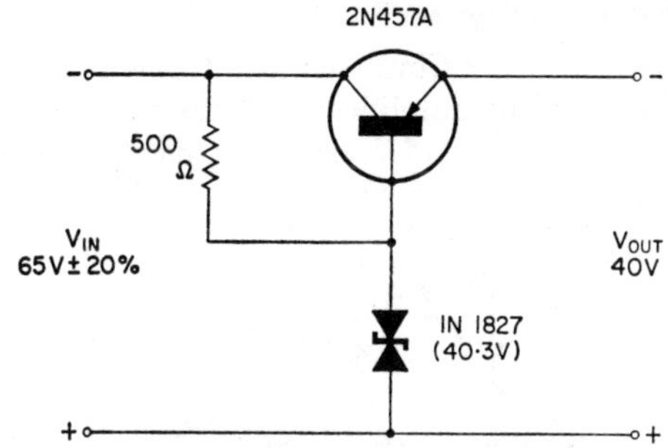

Fig. 15.6.—*Emitter follower stabiliser.*

The base electrode of the transistor is held at a constant potential by means of the zener diode. The output voltage at the emitter of the transistor can only differ from this constant potential by an amount equal to the base–emitter voltage of the transistor, which is only a small fraction of 1 volt for germanium transistors, even over the full range of emitter current.

The excess voltage between the variable input voltage and the constant output voltage will appear across the collector and emitter of the transistor, and so it must be chosen to have a sufficiently high collector voltage rating. Moreover, if the output emitter terminal is inadvertently shorted to earth, the full input voltage will appear across the transistor, and if the collector voltage rating is not high enough to withstand this eventuality, some special protection circuit must be included.

40 Volt 1·5 Ampere emitter follower stabiliser

The values given in Fig. 15.6 refer to an emitter follower stabiliser for 40 volts. A double anode zener diode is used (sometimes called a regulator diode). The circuit has the following performance:

Input voltage	52–78 volts
Output current	0–1·5 amperes
Output voltage	39·2–40·8 volts
Output impedance	0·3 ohms

Feedback series stabilisers

The operation of the simple emitter follower circuit can be improved by including amplified feedback control as shown in the basic circuit of Fig. 15.7. A zener diode *D* is used as a

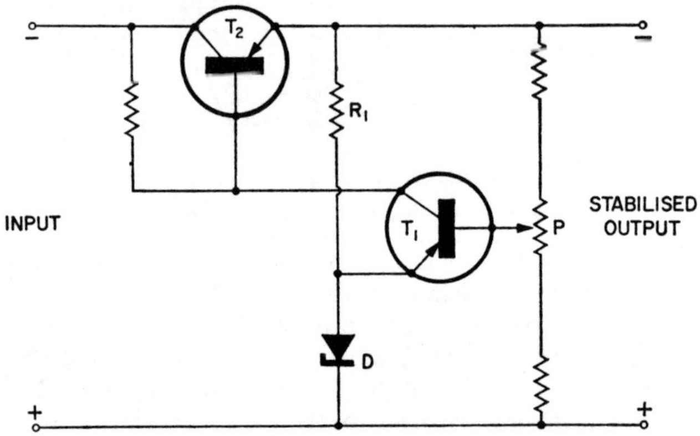

Fig. 15.7.—*Basic series stabiliser circuit.*

constant voltage reference, being biased into the breakdown region through the resistance R_1. The diode holds the emitter of transistor T_1 at the constant reference voltage. Any variation on the output voltage appears at the base of T_1, is amplified, and fed to the base of the series control transistor T_2 resulting in the collector to emitter voltage of T_2 changing to oppose the initial output variation.

Looking at the operation of the circuit in more detail, suppose the output voltage tends to increase, either due to an increase in mains voltage or a decrease of load current. The percentage change in output voltage is fed to the base of the transistor T_1 by means of a potentiometer P across the output voltage. The level of the voltage applied to T_1 is approximately equal to the breakdown voltage of the zener diode connected to the emitter of the transistor. The voltage at the base of T_1 will go more negative, and as the emitter is held constant by the diode D, the emitter–base voltage will increase and cause the collector current of T_1 to increase, and the collector to go more positive. This will reduce the emitter–base bias of the series control transistor T_2 and result in an increase in collector–emitter voltage of that transistor, so that the output voltage is reduced back almost to the original level.

30 Volt 0·4 Ampere stabiliser

A practical series stabiliser circuit for 30 volts 400 mA output is shown in Fig. 15.8. A 6·2 volt reference diode (D_1) is used and the control element consists of a 2N1049 power transistor (actually a Darlington connected pair with a 2N656 transistor (see Chapter 14)). This pair of transistors acts as a 2N1049 with very high gain. The circuit around T_4 and diode D_2 is called a *pre-regulator* circuit, and is included to provide a constant current to the collector of T_3 and the base of T_2. This circuit prevents any ripple existing on the input voltage from appearing at the base of T_2 and being amplified by the series transistor. The capacitor C is included to prevent high frequency instability of the system.

For a change of input voltage of 35–70 volts, the output voltage change is 0·04 volt. For a change in output current from 0 to 400 mA, the output voltage change is 0·05 volt. For a change in ambient temperature from 0°C to 75°C, the output voltage change is 0·2 volt.

50 Volt 2·5 Ampere stabilised supply

A 50 volt stabilised supply is shown in Fig. 15.9. The stabiliser does not use a transistor pre-regulator as in the previous circuit. A resistance–capacitance filter is used to reduce the ripple voltage at the base of the 2N1039 driver transistor. Two 12 volt zener diodes connected in series are used to give the

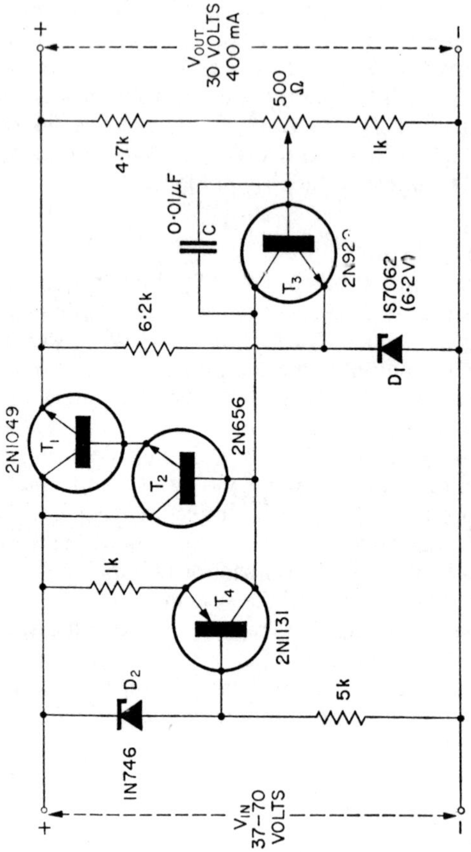

Fig. 15.8.—30-volt, 400 mA series stabiliser.

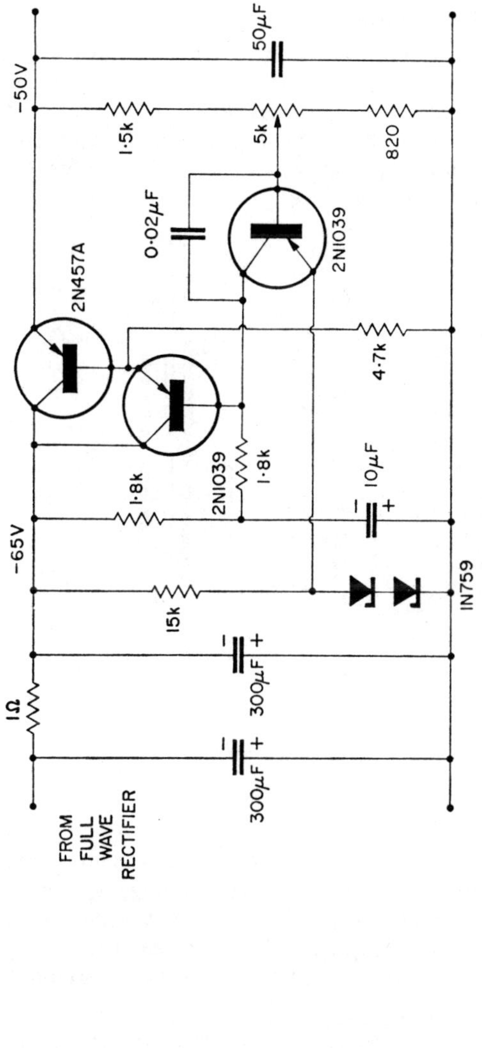

Fig. 15.9.—50 volt stabilised supply.

reference voltage. The circuit is suitable for currents up to 2·5 amps.

D.C. to A.C. inverters

An inverter is a circuit for producing a.c. output power from a d.c. supply. Power transistors can be used in self-oscillating inverter circuits with high efficiency. The basic transistor inverter circuit is shown in Fig. 15.10. It uses two power transistors and a special transformer with a core that has a square-loop hysteresis curve.

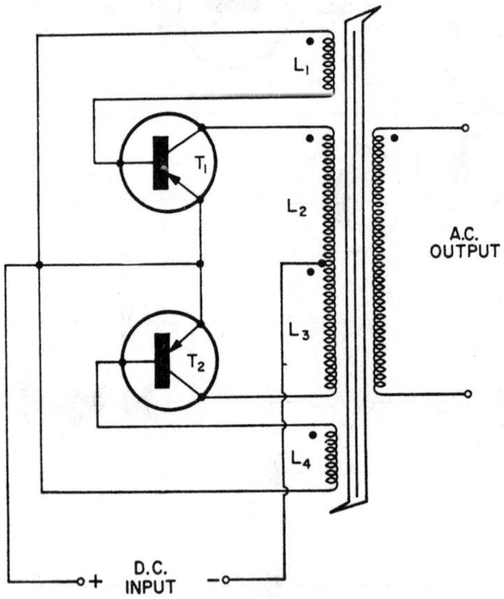

Fig. 15.10.—*Basic transistor inverter circuit.*

When the supply is connected to the circuit, differences in current gain and base to emitter voltage will result in an unbalance so that one transistor conducts more than the other. Suppose that this is transistor T_1. The current through T_1 will flow through winding L_2 and will cause a magnetic flux to be produced in the transformer core. The increasing current will cause an increasing flux which in turn will induce a voltage in the other windings. The sense of the winding L_1 is such that the voltage induced in it is of the correct polarity to drive the transistor T_1 further into conduction, that is, regeneration takes place. At the same time, the sense of winding L_4 is such that the voltage induced in it reduces conduction in transistor T_2. The current in L_2 continues to rise until the transformer coil becomes saturated. Since the voltage induced in the transformer windings is proportional to the rate of change of flux, when the core becomes saturated the induced voltage in L_1 drops to zero and the current through transistor T_1 decreases. The resultant drop of current in L_2 causes the flux to collapse and produce voltages of opposite polarities in windings L_1 and L_4. This reversal of voltage causes transistor T_2 to turn on and transistor T_1 to be held off. The current through L_3 increases until the transformer core becomes saturated in the reverse direction and then the cycle repeats. The output is a square wave, and if a sine wave is required it is necessary to include choke-capacitance filters in the output.

When either of the transistors is not conducting, a voltage approximately equal to the supply voltage is induced across its collector winding, with a polarity that adds to the d.c. supply voltage. Thus each transistor must have a breakdown voltage rating greater than twice the supply voltage.

The operating frequency of the circuit is mainly determined by the transformer design and core material. It will be dependent on the value of supply voltage and to some extent on the value of load current.

If an inverter is required to operate equipment normally fed from the a.c. mains, it will necessary to stabilise the output voltage and frequency. This can be accomplished by including a stabilising circuit in which the collector induced voltage is detected by a sensing circuit, and used to vary the d.c. voltage applied to the oscillator by means of a series regulator circuit. Having produced a stable frequency oscillator, the output voltage and power can be raised to any desired level by means of a power amplifier stage.

Fig. 15.11 shows a complete inverter for an a.c. output of 240 volts, 200 watts at 50 c/s from a 12 volts car battery. Transistors T_3 and T_4 form the oscillator circuit, the induced voltages in the collector windings are rectified by D_1 and D_2

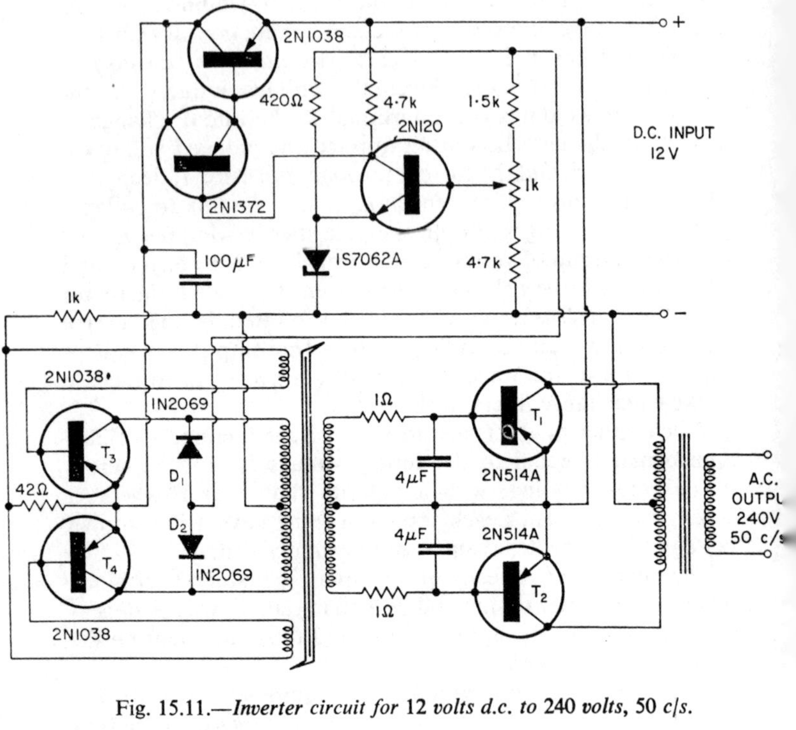

Fig. 15.11.—*Inverter circuit for* 12 *volts d.c. to* 240 *volts*, 50 *c/s.*

to give a d.c. sensing voltage which controls the voltage regulator circuit in the 12 volts supply. This circuit results in the induced voltage being held constant, and this in turn holds the frequency constant. The output from the oscillator is coupled to a power amplifier stage which is fed direct from the 12 volts supply.

The performance characteristics of the inverter are shown in Fig. 15.12.

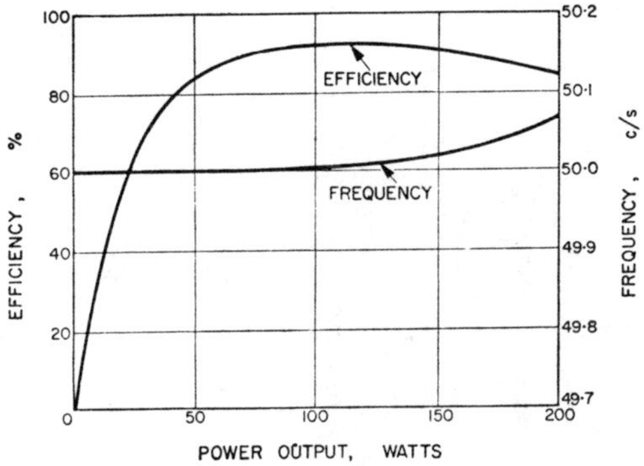

Fig. 15.12.—*Performance of the inverter shown in Fig.* 15.11.

D.C. to D.C. converters

Any d.c. to a.c. inverter can be used as a d.c. to d.c. converter by rectifying the a.c. output. The secondary winding on the oscillator transformer is arranged to give the required voltage level, and, after rectification, suitable smoothing circuits are included. Fig. 15.13 shows the basic d.c. to d.c. converter circuit, the operation of the oscillator being identical to that of the inverter described above. Since with the d.c. to d.c. converter there is no specific frequency requirement (as there may be with a d.c. to a.c. inverter) the operating frequency can be chosen to give optimum performance coupled with small size and economy. Typical values are in the range 400–1,000 c/s. Using the values and devices indicated in Fig. 15.13, an output of 100 volts, 50 watts can be obtained from a 12 volts source.

A typical example of a d.c. to d.c. converter is in a battery operated electronic flash unit for photographic use, as illustrated in Fig. 15.14. The oscillator is fed from a 12 volts battery, and gives a d.c. output of 270 volts which is used to charge a 1,600 μF capacitor. When the trigger contacts are closed by the shutter mechanism on the camera the lamp is triggered and the capacitor discharges through it to give a high intensity flash.

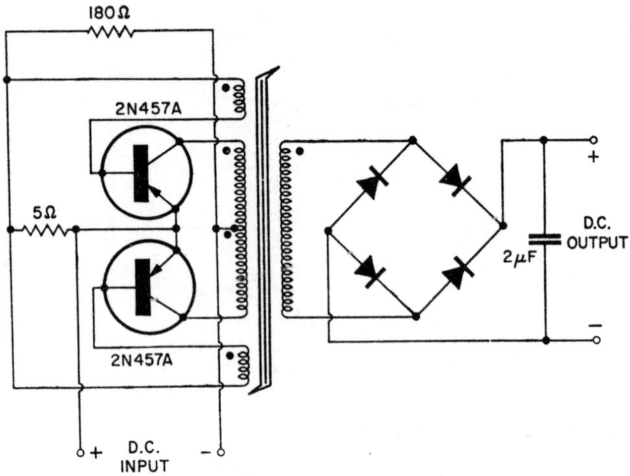

Fig. 15.13.—*Basic d.c. to d.c. converter circuit.*

For high power converters, above about 50 watts, the two transformer circuit shown in Fig. 15.15 may be used. This circuit has the advantage that by using a small saturating transformer TR_1 and a large linear output transformer TR_2, the transformer design is less critical and the arrangement is cheaper than one using a single large saturating transformer. Also, since the current required by TR_1 during saturation is small compared with the load current, a higher percentage of the transistor current rating is used to supply the load, resulting in a higher maximum power output for the same transistors. The magnetising current of the output transformer is low, as it does not saturate.

A small part of the output power is fed back from TR_2 to TR_1 to provide the switching drive to the transistors. As transformer TR_1 reaches saturation, the increasing saturation current causes a voltage drop across the feedback resistance

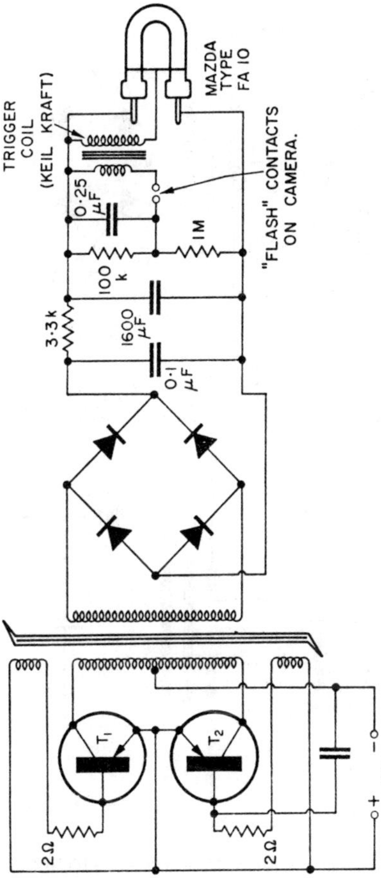

Fig. 15.14.—*Transistor electronic flash unit.*

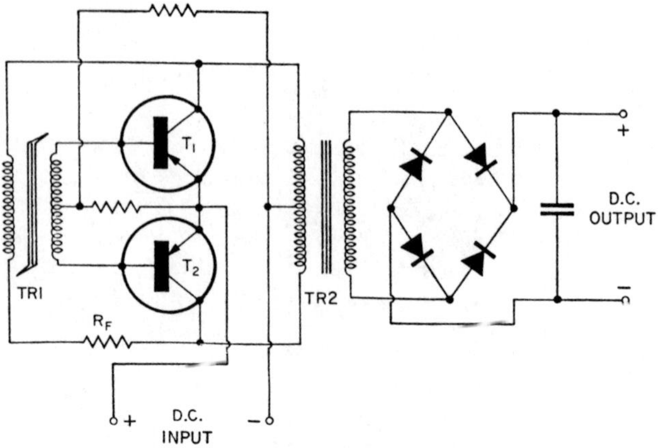

Fig. 15.15.—*Two transformer converter.*

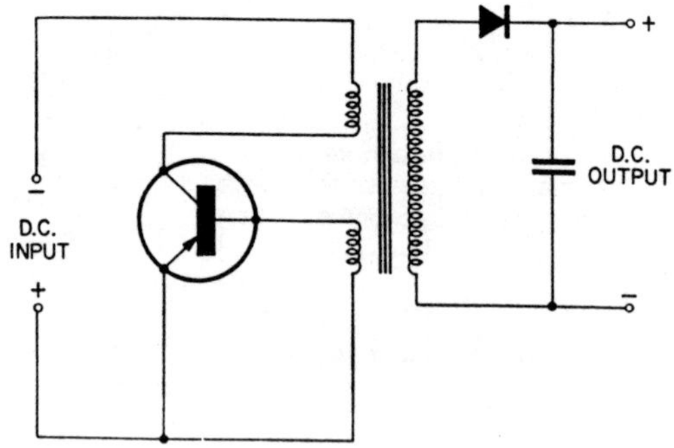

Fig. 15.16.—*Basic ringing choke converter.*

R_F, and results in a decrease in the base drive which turns the conducting transistor off.

Ringing choke circuit

A further circuit sometimes used for converters is the ringing choke circuit shown in Fig. 15.16. This can give an inexpensive and efficient unit for low power applications. The arrangement consists of a transistor oscillator so arranged that energy is built up in an inductance during the period when the transistor is conducting. When the transistor is cut off the inductance "rings" and the energy is taken off through a rectifier to charge the reservoir capacitor.

16
THE USE AND HANDLING OF TRANSISTORS

THE nature and form of transistors require that certain precautions are observed during their handling, installation and use. A booklet entitled *The Use of Semiconductor Devices* has been issued by V.A.S.C.A. (The Valve and Semiconductor Manufacturers Association). This gives good general advice, which has been freely interpreted in this chapter. Due acknowledgement to V.A.S.C.A. is hereby made.

It is always important to ascertain the manufacturer's recommendation on any aspect concerned with the handling and use of his devices. If the published data is not clear on any point, the manufacturer should be consulted—he will doubtless be ready to cooperate.

Electrical ratings

The electrical ratings of a transistor stated by the manufacturer on the data sheet are limiting conditions of operation. If these limits are exceeded by the user, the performance or life of the transistor may be impaired. The important ratings are collector voltage, collector current, collector power and junction temperature. It is necessary for the design engineer to ensure that the transistor will be operated within these electrical ratings, especially during experimental development work. It must not be assumed that one rating, say collector current, may be exceeded if a corresponding reduction is made on another rating—e.g. collector voltage. The manufacturer usually states the ratings as "absolute maximum" values, that is, they must not be exceeded under any circumstance. He ensures, in the preparation of the ratings, that the device will operate satisfactorily at all the limiting ratings simultaneously.

Thermal considerations

With regard to temperature ratings there are two different aspects. It has been shown that the performance of a transistor eventually falls off as the temperature is increased. The maximum operating junction temperature for germanium devices is in the region of 85°C and for silicon transistors about 200°C. When not operating, it may be possible that transistors can be subjected to higher temperatures without damage. The limiting temperature from the permanent damage viewpoint depends upon the materials used in the construction. With the germanium alloy transistor, the limit is the melting point of indium, which is used to form the junctions, and with silicon mesa and planar transistors the limit is the melting point of gold–silicon eutectic which is formed when the silicon is fused to the gold plated header. The manufacturer usually quotes a maximum "storage" temperature: this temperature is higher than the maximum operating temperature, and it is most important not to exceed it during installation. Thus when transistors are soldered into circuit care must be taken to prevent excessive heat flow up the leads into the body, since the thermal capacity of the transistor element is quite small and it is relatively easy to cause an excessive temperature increase. For the soldering operation the usual precautions should be observed—a clean soldering iron bit and cleaning and pretinning the transistor leads and soldering tags. A small-type soldering iron should be used, and the practice of introducing a thermal shunt between the soldering point and the transistor is strongly recommended. This can conveniently be done by gripping the wire with a pair of pointed nose pliers, or by clipping on a spring loaded clamp. The soldering operation should always be completed as quickly as possible.

Handling during installation

Before starting the installation of a transistor, reference should be made to the manufacturer's data sheet to ascertain or confirm the electrode connections. Careful attention must be paid to the type of transistor—pnp or npn—so that the correct polarity can be arranged for the power supplies and bias circuits. For exploratory development work, the use of transistor plug-in sockets is recommended. This avoids the need for continual soldering and allows flexibility in experiments. In many cases the transistor leads need not be cut short at this stage, thus avoiding possible wastage due to leads being too short for other applications.

It is very important that the leads are not bent too close to the transistor, as damage could be caused to the glass–metal seals and allow ingress of moisture. A distance of 0·15 inch should be allowed from the base before bending a wire. Small spacers which thread over the wires and locate up to the base of the transistor are commercially available. These prevent intentional or accidental bending of the wires close to the glass seal.

With stud or flange mounted power transistors it is important to ensure that they are mounted tightly on to the heat sink, but care should be taken to avoid the application of excessive force, otherwise mechanical damage to the unit may result. In this connection it is especially important to ensure strict compliance with the manufacturers recommendations, since the overall power rating of the device plus heat sink will depend on the assembly.

When connecting a transistor into circuit, either by plugging into a socket or soldering direct into circuit, the power supplies must be switched off. Apart from not having the possibility to cross-check the connections after the transistor is in circuit and before power is applied, it is possible that damage could result to the transistor from the discharge of capacitors or other effects. When transistors are being soldered into circuit, care should be taken to ensure that the soldering iron is earthed. If it is not, and the insulation between the heater winding and case is poor, it is possible that damage could be caused to the transistor by leakage currents from the case.

Wherever possible, it is advisable to mount the transistor so that some air convection current flow is possible, especially when the device is required to dissipate power approaching the rated value.

Circuit measurements

In view of the small size of transistors and the small spacing between the connections, great care must be exercised when making measurements while the circuit is operating to avoid an accidental short circuit. This is particularly the case when printed circuits are used. An accidental connection from the base of a transistor to the collector supply line, or a short circuit across an emitter resistor, could well cause sufficient current flow in the transistor to cause permanent deterioration. Well made, insulated probe electrodes should be used with the measuring instrument. For voltage measurements, the instrument should have a resistance of at least 20,000 ohms per volt with full scale ranges of 1, 10 and 100 volts.

Perhaps the most useful of all measurements on an operating circuit is the d.c. voltage across the emitter resistance of a transistor. A satisfactory reading will indicate that the stage is probably operating correctly.

It is possible in some cases to measure the value of a resistor while in circuit, but the result may need careful attention to ensure correct interpretation. If a convential ohmmeter is used, the internal voltage should not exceed 1·5 volts.

General circuit considerations

While economic considerations often dictate that the ratings and capabilities of a transistor are fully utilised, it is prudent for the user to establish a margin of safety in the circuit design for any particular application. If reliability is more important than cost, it is sensible to limit the junction temperature by restricting the power dissipation to some fraction of the quoted rating. Similarly if there is the possibility of voltage transients, the transistor should be selected to anticipate these, even though the operating d.c. voltage may only be a small fraction of the maximum voltage rating. It may be possible of course to devise circuits to limit the value of the transients; the choice would eventually depend upon economic considerations. Whenever possible, circuits should be designed to include suitable limiting features so that failure of an associated component will not result in overloading of the transistors.

Simple transistor tester

Before putting a transistor into circuit, it may be desirable to be able to carry out a rough check to confirm that it is generally satisfactory. A simple circuit for checking gain is illustrated in Fig. 16.1. Separate sockets are provided for pnp

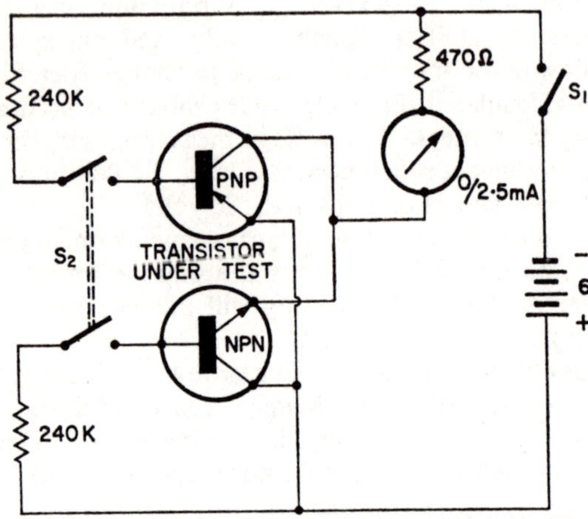

Fig. 16.1.—*Simple transistor tester.*

and npn transistors. The transistor to be tested is plugged into the appropriate socket and switch S_1 closed. The base electrode of the transistor is open circuit, and so the meter in the collector circuit will read I_{CEO}. With a good transistor only a very small reading will be obtained on the meter. When the spring loaded switch S_2 is depressed, a current of 25 μA is fed into the base, and if the meter (0–2·5 mA) is calibrated with a full scale gain of 100, the transistor gain can be read off directly.

17
SOLID CIRCUIT TECHNIQUES

INTEGRATED CIRCUITS

SINCE the early days of electronics, there have been continual efforts to miniaturise electronic equipment by making individual components smaller and by devising more concise assembly techniques using printed circuits, potted modules and the like. This general trend has culminated in what is known as *microelectronics* and more specifically from the semiconductor viewpoint as *integrated circuits.* By an integrated circuit, we mean a circuit which has been manufactured as an inseparable assembly of electronic parts, in a single structure which cannot be divided without destroying its intended electronic function. At the present time there are four distinct approaches to integrated circuits.

(1) *Thin film circuits.* In this approach, passive components —resistors and capacitors—are prepared by vacuum deposition of metals and dielectrics on to a glass or ceramic substrate. The active components—transistors and diodes—are prepared as miniature individual units in the usual way, mounted in suitable locations on the substrate and connected to the passive elements by thermal compression bonding of small gold wires.

(2) *Solid circuits.** In this method, all components, both active and passive, are formed in a single wafer or block of semiconductor material—usually silicon—by a diffusion process. Interconnections between components are formed either internally during diffusion, or by means of metallic paths deposited on to the oxidised surface of the wafer.

(3) *Multiple chip circuits.* Here the passive and active components are formed in separate wafers, or "chips", of semiconductor material, which are then mounted on to a common base and interconnected by thermocompression bonding.

* "Solid circuit" is a registered trade mark of Texas Instruments Inc.

(4) *Hybrid integrated circuits*. This is a combination of the first and second approaches, in which the active components are first prepared in a single wafer of silicon with an oxidised surface, and the passive components formed on the surface of the oxide by vacuum deposition.

The first method, the thin film circuit, uses techniques that have been established for some time. It is essentially an extension of conventional electronics in that it can be partly considered as an assembly of discrete components, retaining a measure of flexibility which has resulted in short development times and has allowed fabrication with reasonable initial cost.

The technology of solid circuits has been derived from that used to make silicon planar transistors, and considerable progress has been made in this field. There can now be little doubt that the solid circuit method will, in the relatively near future, offer advantage in all aspects—small size, economics, reliability and performance. The whole field of electronics is now opening up to the design and application of these complete circuit functions.

The topic of solid circuit application is widening rapidly and is too great for the scope of this book, but a survey of the methods used in the fabrication of solid circuits will be given to allow a basic understanding of the subject.

SOLID CIRCUITS

A solid circuit is a complete electronic circuit, comprising transistors, diodes, resistors and capacitors, together with the necessary interconnections, completely fabricated in a single small wafer of semiconductor material. To give an indication of the possibilities of the technique, solid circuits are being manufactured at the present time in which 38 transistors and 31 other components are formed and interconnected in a wafer of silicon 0·18 inch long by 0·06 inch wide by 0·010 inch thick. The various components are all formed in the wafer of silicon using the diffusion–planar process as described for planar transistors in Chapter 4. The initial silicon slice has a relatively high resistivity and by means of a pn junction each component is electrically isolated. The dimensions and location of the diffused components are accurately determined by the photomasking process, using a silicon dioxide coating on the surface of the slice. Silicon dioxide is an effective mask against diffusion and so, by selectively removing the oxide from the surface, diffused areas can be formed where the oxide has been removed. During the diffusion process, silicon dioxide is reformed over the surface of the diffused areas. This reformed oxide is now selectively removed and ohmic contacts made to the diffused areas by depositing aluminium. At the same time aluminium paths are deposited over the insulating silicon dioxide surface to provide point to point connections between the various components so as to give the required electronic circuit.

As with planar transistors, a number of identical units can be fabricated simultaneously on a single slice of silicon. On a slice approximately 1 inch diameter, 40 of the solid circuits mentioned above can be made. After the simultaneous processing has been completed, the slice is cut into individual circuits by a scribing operation, and each circuit fixed to a special mount and hermetically sealed.

The solid circuits fabricated in this way are not only extremely small, there can be little doubt that they will be much more economic and more reliable than the conventional discrete component equivalents.

SOLID CIRCUIT COMPONENT FORMATION

Electrical isolation

Since all the components are formed in a piece of conducting silicon, it is necessary to arrange that they are electrically isolated. To do this, a p-type substrate slice is used, and the first diffusion of each component made n type. The pn junction so formed is biased in the reverse direction and so presents a high resistance. Thus between any two components there will be two reverse biased junctions in series, providing high resistance isolation. The nature of the isolation will be considered further during the discussion on the formation of the various components.

Transistors

The techniques for making transistors in solid circuit form are similar to those used in the fabrication of normal planar

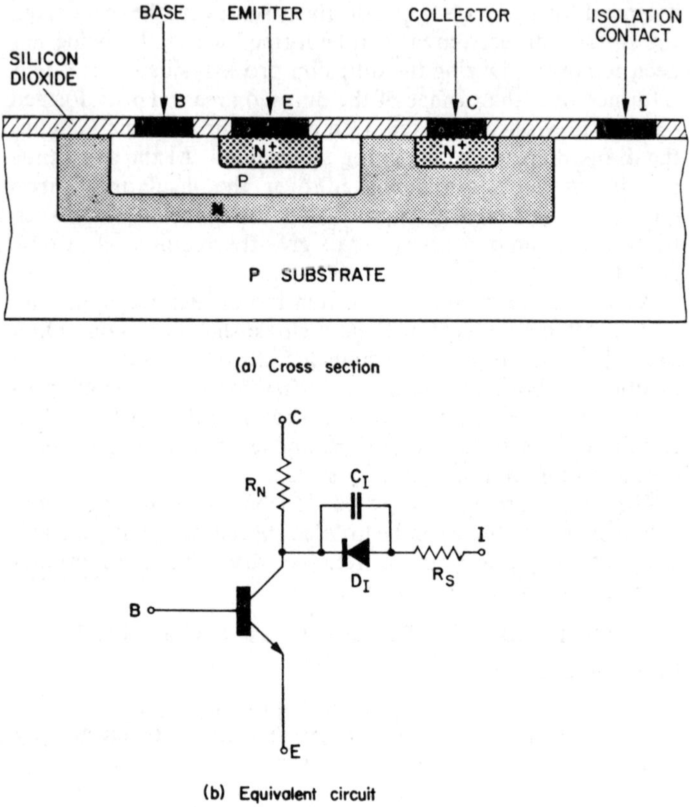

Fig. 17.1.—*Solid circuit planar transistor.*

transistors, as described in Chapter 4. A cross-sectional view of a solid circuit transistor is shown in Fig. 17.1 (a). The n-type collector region is first formed by diffusing phosphorus into the p-type slice. This is followed by diffusing boron to form the p-type base region and finally a second phosphorus diffusion to form the n+ emitter. At the same time as forming the emitter, an n+ region is diffused into the n-type collector region so that a low resistance collector contact can be made

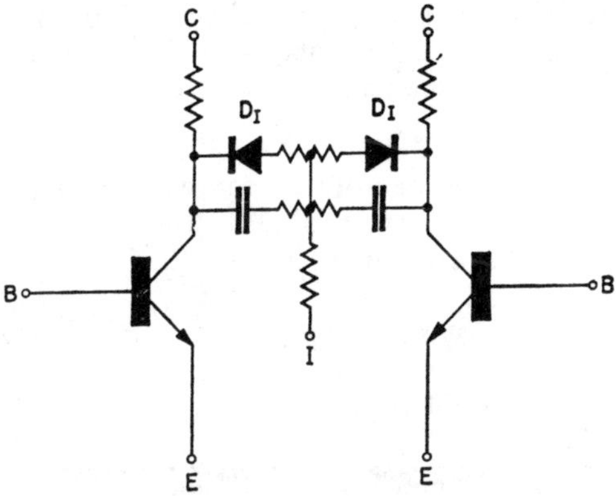

Fig. 17.2.—*Two transistor equivalent circuit.*

on the top surface. The top collector contact is necessary to allow the interconnections to be made, but it introduces an increase in the series collector resistance as the collector current must flow along the narrow n-type collector region to reach the contact. This results in a small increase of the collector saturation voltage $V_{CE(sat)}$ over that of the conventional planar transistor.

The equivalent circuit of the solid circuit transistor, including the "isolation diode", is shown in Fig. 17.1 (b). The isolation diode D_I will have a junction capacitance C_I, shown in parallel, and the series resistance R_S is due to the bulk resistance of the substrate region between the isolation junction and the isolation terminal. If the latter is connected to a point slightly negative to the collector potential, the diode D_I will be reverse biased, and so will present a high isolation resistance. Fig. 17.2 shows the equivalent circuit of two solid circuit transistors formed side by side. It will be seen that the two isolation diodes appear back to back between the two collectors. Thus even if the isolation terminal is not connected to any reference potential the two collectors will be isolated from each other since one of the diodes will always be reverse biased. However the substrate is usually connected to the negative supply terminal so that each diode will be reverse biased. With this arrangement it is important to remember that the diode capacitance may result in a low isolation for a.c. or transient signals.

Diodes

PN junctions can be formed for use as diodes during the transistor diffusion processes. Two types are used. Fig. 17.3 illustrates a collector–base diode. The n-type cathode

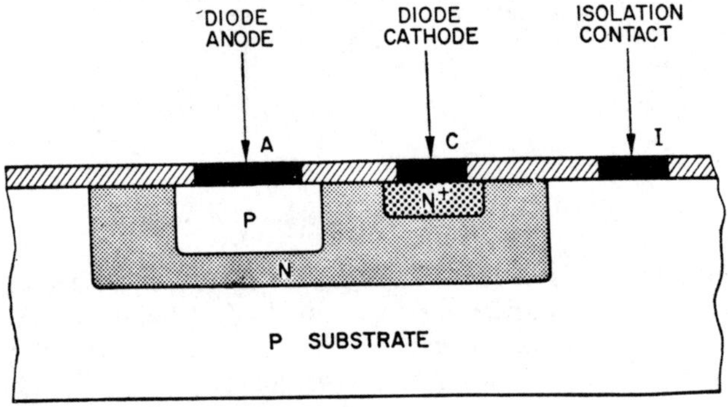

(a) Cross section

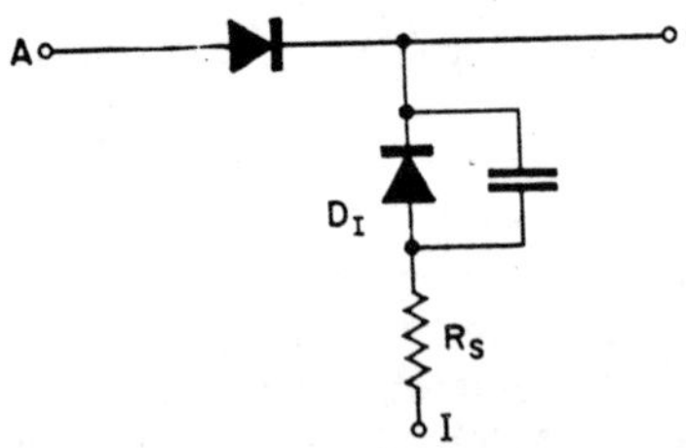

(b) Equivalent circuit

Fig. 17.3.—*Solid circuit collector-base diode.*

region is formed during the collector phosphorus diffusion, the p-type anode region is formed during the boron base diffusion, and the n+ cathode contact area during the emitter

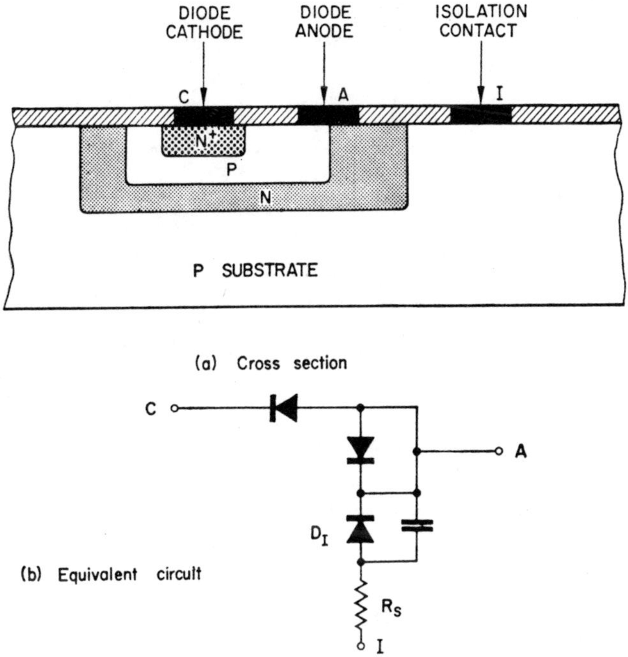

Fig. 17.4.—*Solid circuit emitter-base diode.*

diffusion. The equivalent circuit shows the isolation diode which is similar to that for the transistor. The second type is the emitter–base diode illustrated in Fig. 17.4. An n+pn structure is formed similar to a transistor. The n+ emitter region forms the diode cathode, and the p-type base region the diode anode. To avoid spurious effects due to transistor action, the p-type base and n-type collector regions are short-circuited by the anode contact. Again the isolation is similar to that for the transistor. Emitter–base diodes offer advantage for fast switching purposes, but their breakdown voltage is less than that of collector–base diodes.

Resistors

A resistor can be formed in a wafer of high resistivity silicon by the diffusion of an impurity into a defined region. The

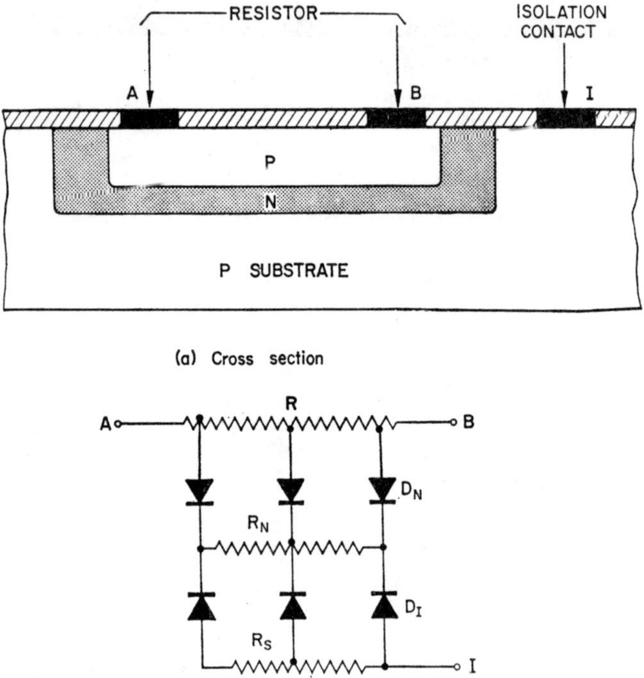

(b) Equivalent circuit

Fig. 17.5.—*Solid circuit resistor.*

value of resistance will depend upon the dimensions of the region, the surface concentration of the diffusant element, and the diffusion depth. For economic reasons, it will normally be desired that resistors are formed during one of the transistor diffusion processes. The surface concentration and diffusion depth will be fixed by the requirements of the transistor, and so the dimensions of the resistor region must be designed to give the correct value of resistance. Values of resistance between 20 and 20,000 ohms can be obtained using the p-type base diffusion. This method is illustrated in Fig. 17.5 (a) and the equivalent circuit is shown in Fig. 17.5 (b). The main resistance R is formed in the p-type region. The resistance of the n-type region R_N does not appear in parallel with R because as soon as there is any current through R causing a voltage drop between the ends one of the equivalent diodes D_N will be reverse biased. In any case, the resistance of the n-type region is much higher than that of the p-type region, which has a high-concentration of carriers near the surface of the wafer. The isolation diodes D_I are reverse biased by connecting the substrate to the negative supply.

An important feature of the formation of resistors is that it is possible to form two resistors side by side with a very precise ratio of their resistance values. Although it is difficult to reproduce individual resistors to closer than ± 10 per cent, with two resistors side by side any variation will affect both resistors equally and the ratio can be reduced to ± 1 per cent. Moreover, the variation of resistance with temperature will be the same with each resistor. This precise ratio can be used to advantage in circuit design.

If resistance values smaller than about 20 ohms are required, the emitter diffusion can be used, and the resistor formed in the n+ region.

Capacitors

Capacitors can be obtained by using the depletion capacitance of a pn junction. This has the limitation that the value of capacitance is dependent upon the voltage across the junction, but the junction can be formed at the same time as the transistors, diodes and resistors. In order to obtain the maximum value of capacitance per unit area, a structure similar to a transistor is formed, and the two junctions connected in parallel as shown in Fig. 17.6. The n+ and n regions are connected together and form one terminal, the p-type region being the other terminal. Isolation is again arranged as for the transistor by the junction between the p-type substrate and the n-type region.

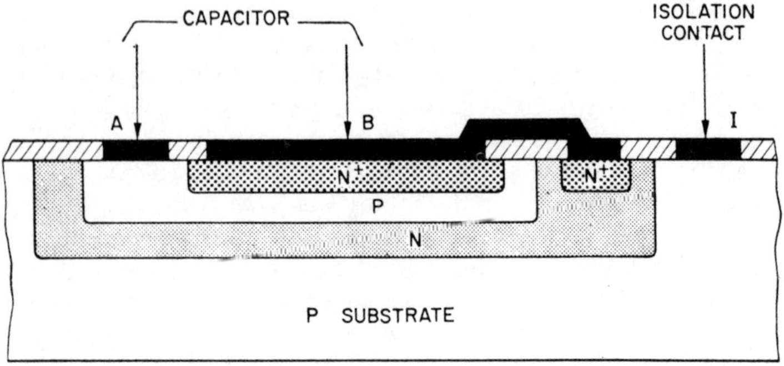

(a) Cross section

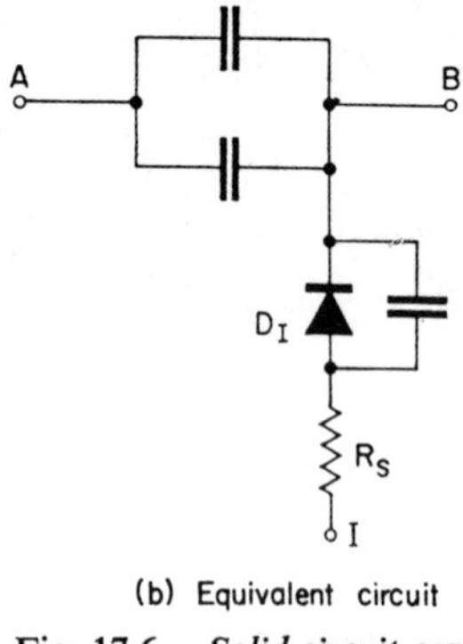

(b) Equivalent circuit

Fig. 17.6.—*Solid circuit capacitor.*

An alternative method of forming a capacitor is to use the silicon dioxide on the surface as a dielectric by evaporating a metallised electrode on to the oxide in the required region.

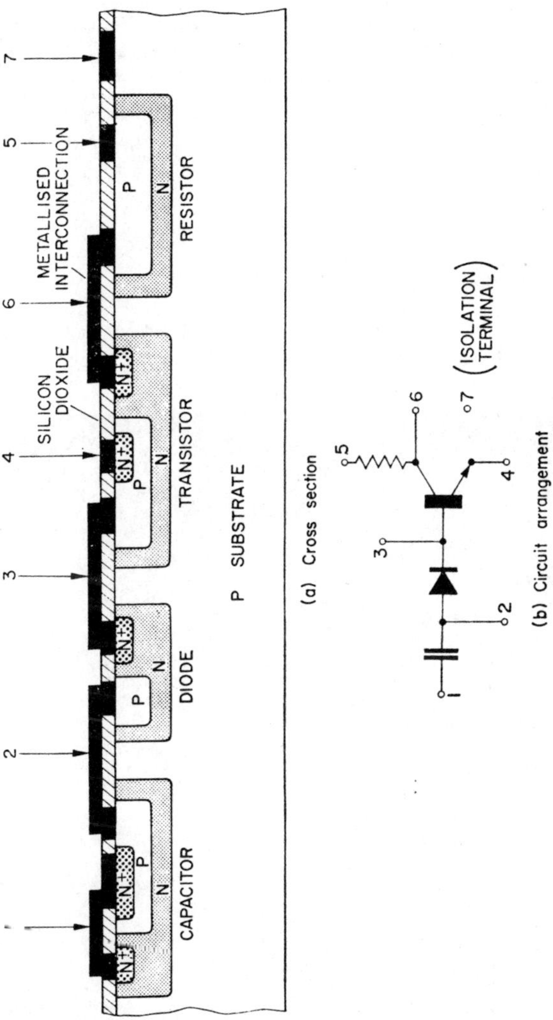

Fig. 17.7.—Part of solid circuit illustrating interconnections.

Circuit formation

The formation of various components into a complete circuit is illustrated by the typical combination shown in Fig. 17.7. The components are interconnected by evaporating metal connections over the silicon oxide as shown. The illustration is, of course, only two dimensional. Further components can be connected in a similar manner in the third dimension.

Circuit fabrication

The steps in the manufacture of a solid circuit are illustrated in Figs. 17.8 and 17.9. Fig. 17.8 shows a completely processed

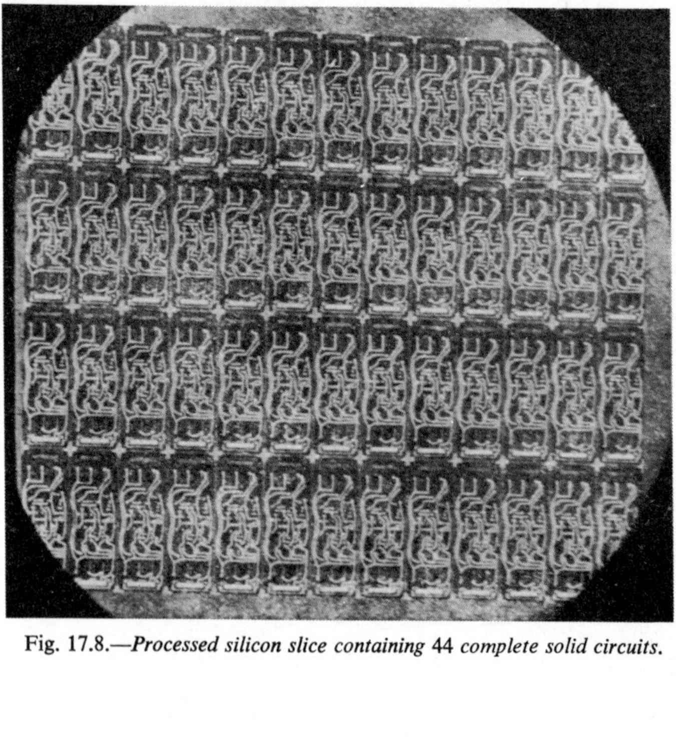

Fig. 17.8.—*Processed silicon slice containing* 44 *complete solid circuits.*

slice containing 44 solid circuits, ready for scribing into individual units. Up to this point all processes have been carried

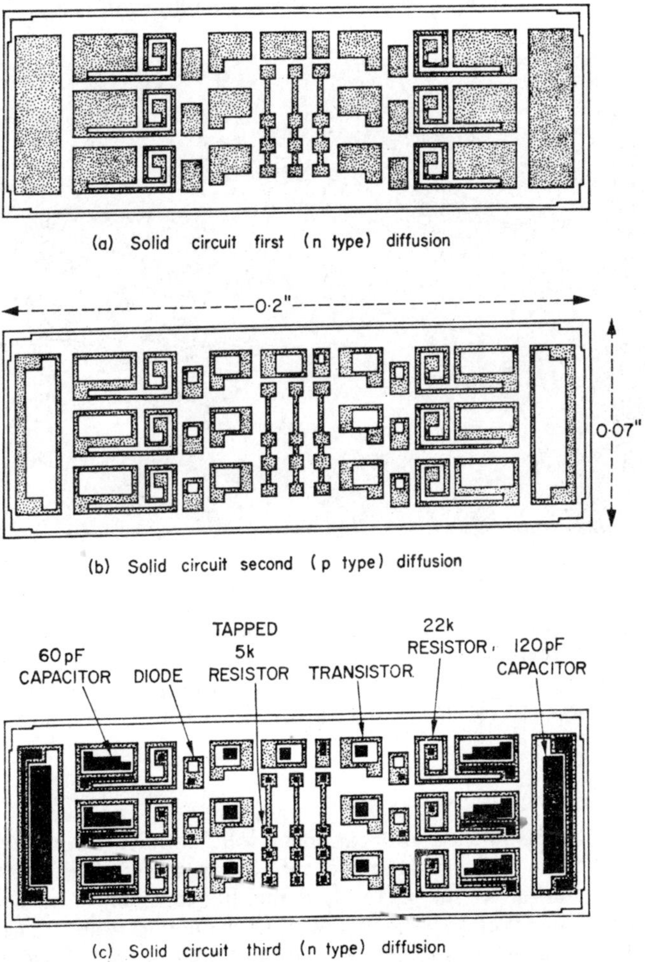

Fig. 17.9.—*Diffusion stages in solid circuit preparation.*

out on the whole slice. Fig. 17.9 (a) shows one of the circuits on a slice after the first, n-type, diffusion. Fig. 17.9 (b) shows the same circuit after the second diffusion, the p-type base diffusion, and Fig. 17.9 (c) after the n+ emitter diffusion. On

Fig. 17.10.—*Solid circuit mounted in case.*

Fig. 17.9 (c) various individual components have been identified. Fig. 17.10 shows a single circuit assembled into its case ready for final sealing. The overall dimensions of the case are $\frac{1}{4}$ in. × $\frac{1}{8}$ in. Fig. 17.11 shows the electronic circuit of the unit in Fig. 17.10.

Circuit design trends

In the design and application of solid circuits, the relative advantages and disadvantages of different circuit elements, with relation to the method of manufacture, must be considered in order to achieve the most economic result. There are certain features of the technology that will tend to dictate the circuit arrangement.

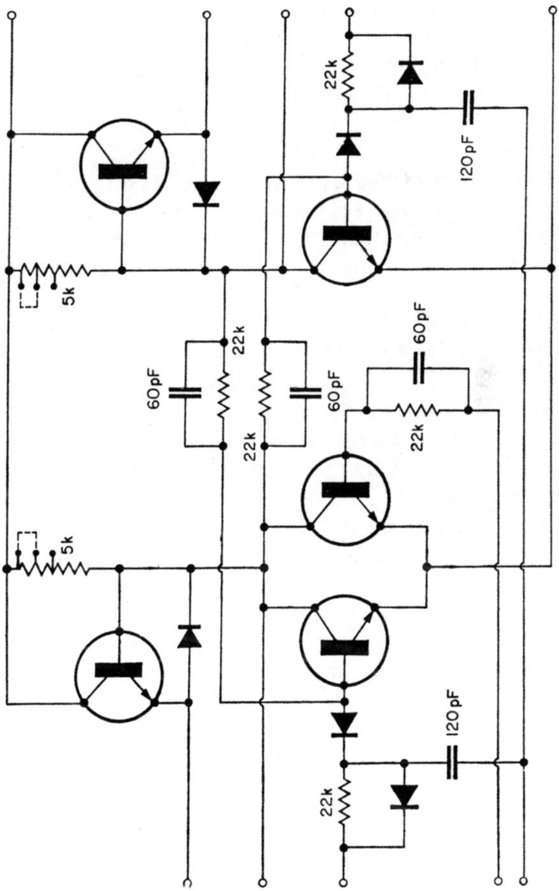

Fig. 17.11.—*Circuit diagram of the bistable solid circuit shown in Fig.* 17.9.

With solid circuits the processing cost will be roughly proportional to wafer area, and so for lowest cost circuit elements of smallest area must be chosen. Resistors and capacitors made in solid circuit form take up considerably more area than transistors and diodes, and so active components become preferable to passive components, which is a complete reversal from previous electronic circuit design. Thus direct coupled circuits with a minimum of resistors will offer the lowest cost circuits.

If the electrodes of two or more transistors or diodes can be made common, considerable saving in area can be made since only a single isolation region will be needed. This can be quite important in logic circuits where there may be multiple input or output circuits.

It was mentioned earlier that the control of resistor values in solid circuit manufacture is not high—about ±10 per cent, but that resistor combinations with a precise ratio are possible to ±1 per cent. Thus circuit design will tend to avoid circuits relying on the precise absolute value of a resistor and where possible will take advantage of the closely controllable ratio between two or more resistors.

These few examples will have served to indicate that circuits devised for the application of solid circuits may differ considerably in approach from the old discrete component electronic circuits. We are now at the start of a new technology which, undoubtedly, will result in applications considered quite unachievable even at the present time.

18

RECENT DEVELOPMENTS

WHILE the major research and development effort in the semiconductor field is probably being directed to integrated circuits, research teams are continually striving to create new discrete devices which will offer improvement in performance, or will open up some new field of application. A few of the developments that have been reported over the last year or two are described in this final chapter.

Metal-oxide semiconductor transistor

In the original type of field effect transistor, the conductivity of a region of semiconductor is *reduced* by increasing the reverse bias across a pn junction so that the depletion layer associated with the junction moves into the region and

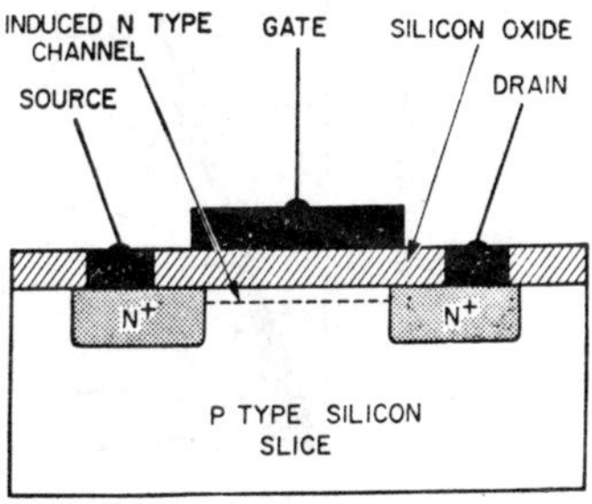

Fig. 18.1.—*Construction of the metal-oxide semiconductor transistor.*

effectively reduces its width. In the metal-oxide semiconductor transistor—originally called the insulated-gate transistor—the conductivity of a region of semiconductor is *increased* by inducing a conducting channel between two electrodes. The channel is induced by increasing the electric field between an insulated gate electrode and the semiconductor region. The arrangement is illustrated in Fig. 18.1. The two electrodes, source and drain, consist of high concentration n-type regions diffused into a p-type silicon slice with very small spacing, about half a thousandth of an inch. A thin layer of silicon oxide, about 1,000 Ångströms, is grown on the surface of the region between the source and drain, and an aluminium gate electrode is deposited on top of the oxide.

With a voltage applied between source and drain, only a very small current flows as there are two pn junctions back to back. If a positive potential is applied to the gate electrode, a negative charge is induced at the surface of the slice under the oxide, forming an n-type conducting channel, and so a majority current will now flow between source and drain, the magnitude of the current being controlled by the potential of the gate electrode. Fig. 18.2 shows the characteristics of a typical

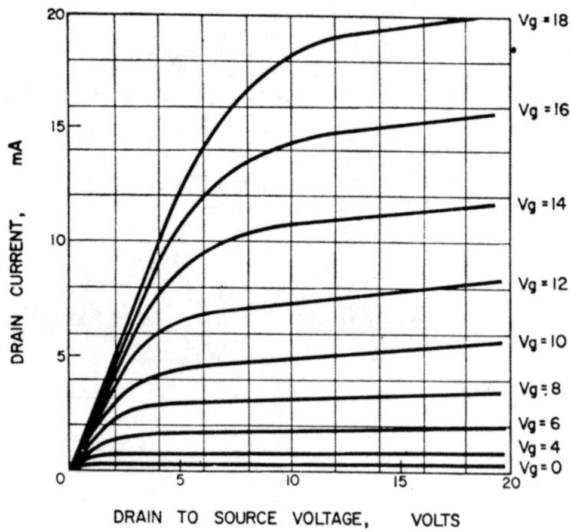

Fig. 18.2.—*Metal-oxide semiconductor transistor characteristics.*

insulated gate transistor, illustrating the effect of gate voltage on the drain current.

Since the gate electrode is formed on a layer of silicon oxide, the gate input impedance is extremely high—up to the order of a million Megohms. Thus the device is complementary to the conventional bipolar transistor with its low input impedance. It will obviously find wide application where high input impedance is required, and may also be particularly useful in switching and direct coupled logic circuits.

Thin film transistors

The thin film transistor is an active circuit element fabricated by evaporating thin films of certain elements on to an insulating substrate. The main objective of developing a thin film

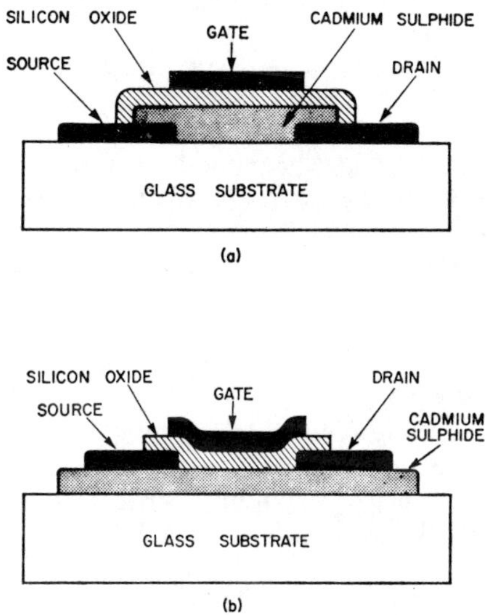

Fig. 18.3.—*Thin film transistor arrangements.*

transistor is so that both active and passive components can be fabricated simultaneously by evaporation, for the manufacture of thin film circuits. The geometry of the thin film transistor is basically similar to that of the insulated gate transistor as illustrated in Fig. 18.3 (a). The insulating substrate is optically polished glass, and source and drain contact electrodes are first formed by evaporating gold, with a spacing in the region of a half to one-thousandth of an inch. The active layer is cadmium sulphide, which is evaporated to a thickness of about 1,000 Ångströms. Silicon monoxide is evaporated on top of the cadmium sulphide to form an insulation layer with thickness between 1,000 and 4,000 Ångströms, and finally the gate electrode is evaporated on top of the oxide over the region between the source and drain electrodes. It has been put forward that the operation of this type of transistor is similar to that of the metal-oxide semiconductor transistor, in that the characteristics result from changes in the conductivity of a channel induced in the cadmium sulphide between the source and drain.

A modified structure to facilitate manufacture is shown in Fig. 18.3 (b). In this arrangement, the cadmium sulphide is first evaporated direct on to the substrate. Then the source and drain electrodes are evaporated on top of the cadmium sulphide, followed by evaporation of the silicon oxide insulation layer and the gate electrode. There are still problems to be overcome on this device and development work is being continued.

Light emitting diode

The light emitting diode—or semiconductor lamp, as it is sometimes called—is essentially a pn junction operated with forward bias. When the minority carriers, which have been injected across the pn junction, recombine, a certain amount of energy is released. This energy can appear in a number of ways, one of which is the emission of light. The amount of light and its wavelength depend upon the characteristics of the materials used to make the pn junction. In practice, intermetallic compounds of groups III and V elements give best light emission. Typical compounds used are gallium arsenide and gallium phosphide. By doping the compound with other elements the wavelength of the light emission can be changed. For example, gallium phosphide doped with silicon and zinc gives a green light, and doped with oxygen and zinc gives a red light.

A typical light emitting diode is the Texas Instruments type SNX100. This uses a gallium arsenide wafer with a pn junction produced by diffusing in zinc, and gives an infra-red output. The size of the wafer is 0·05 inch square by 0·005 inch thick, and the light is emitted from the top surface. With a forward bias of approximately one volt, 100 mA, the light output at room temperature is of the order of 50 microwatts of light power. Higher output power can be obtained by operating with a lower ambient temperature—about 500 microwatts at liquid air temperature. In a more recent development, the flat wafer has been changed to a dome-shaped wafer to give higher efficiency of light emission.

The generation of light depends upon recombination, which is a fast process, so that the light emitting diode avoids the speed limitations of other lamps, such as the heating of a filament or the ionisation of a gas. The type SNX100 diode can be modulated with frequencies up to several Mc/s.

Typical applications for semiconductor light diodes are in short range optical communication systems, including television transmission, as miniature lamps for card readers in computers, and as monochromatic light sources in spectroscopy.

Semiconductor relay

An interesting application of the light emitting diode is its use in conjunction with a silicon photodetector to form a four terminal device with the output completely isolated from the input. The system is similar to a relay in that an actuating signal applied to the input will change the impedance between the output terminals from high to low. The arrangement is illustrated in Fig. 18.4. A silicon photodetector—either a photodiode or a phototransistor—is mounted in close proximity to the light emitting diode and the assembly encapsulated in a transistor case. An input signal applied to the light emitting diode causes a light output which falls on the photodetector and results in an output signal. Although experimental at the present time, there would appear little doubt that it will eventually be established as a standard device.

The use of light coupling also offers the possibility of siting the two elements apart in different sections of an equipment, and coupling by means of a flexible "light pipe".

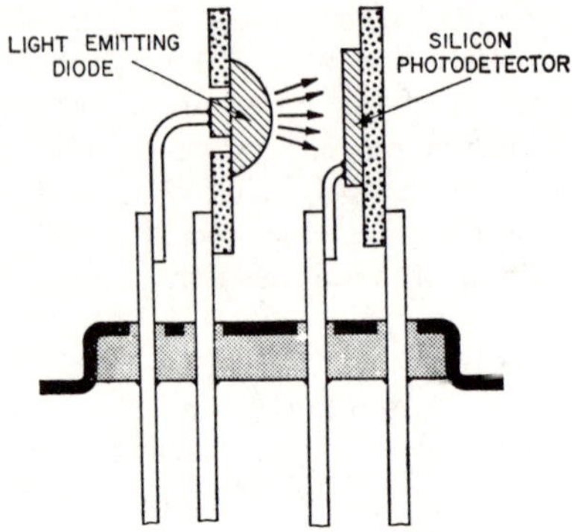

Fig. 18.4.—*Semiconductor relay.*

Semiconductor laser

In a light emitting diode as described above the light output is non-coherent, that is, as holes and electrons recombine the light signals produced over the pn junction area are random in time, direction and polarisation. Most of these signals are emitted in directions which take them away from the junction to form the light output from the top surface of the wafer. However, some light will move along the plane of the junction, and will trigger off additional light signals by a process called stimulated emission. The stimulated signals will be exact replicas of the triggering signals and so they also will move along the plane of the junction, producing still more light. Thus the initial light signal moving along the junction can be multiplied manyfold before it reaches the surface. If now some of the light is reflected at the surface, travels back to the other surface and is partially reflected there, so that it "bounces" to and fro, it will be amplified further with each bounce, and coherent light output will be produced, with the radiation concentrated from the junction region. This is laser action, and the diode is then called a *semiconductor laser*—or a *junction laser*.

The basic form of semiconductor laser described by R. N. Hall (U.S.A.) is shown in Fig. 18.5. The pn junction is formed

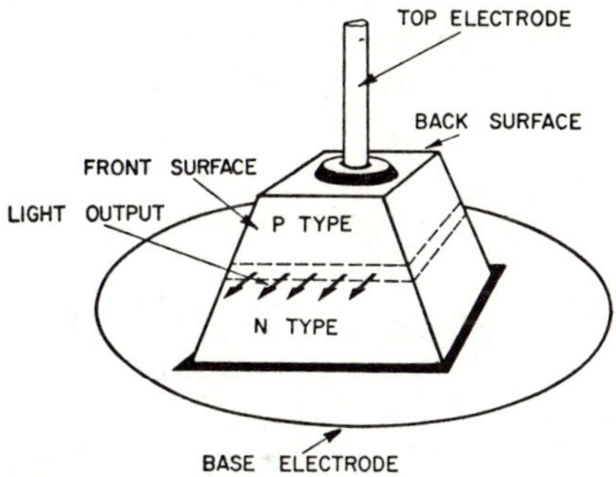

Fig. 18.5.—*Semiconductor laser. Note that the front and back surfaces must be parallel.*

by diffusing zinc into n-type gallium arsenide. A wafer is cut from the slice and the front and back faces precisely ground and polished to be parallel to each other and perpendicular to the junction. The reflectivity of the gallium arsenide to air interface is sufficient to give the required reflection, and so no further surface treatment is needed. The other two faces of the wafer are intentionally made irregular to give non-consistent reflection. The distance between the two reflecting faces will determine the frequency at which the laser will operate—the radiation must have a wavelength such that the distance between the two faces is equal to an integral number of half wavelengths. Typical laser wafers are between 0·2 and 0·4 cm square at the junction.

With such a semiconductor laser, a current density of several thousand amperes per square is necessary to initiate laser action at liquid air temperature, and the light output is concentrated over a very small area in the region of the junction—about 0·005 sq mm—with a peak intensity of the order of 10^6 watts per sq cm.

Semiconductor microphone

It has been known for some time that mechanical stress alters the characteristics of a pn junction. Two methods have been proposed to utilise this effect in a semiconductor microphone. The first effect uses a pn junction diode with local pressure applied with a stylus as shown in Fig. 18.6 (a). With

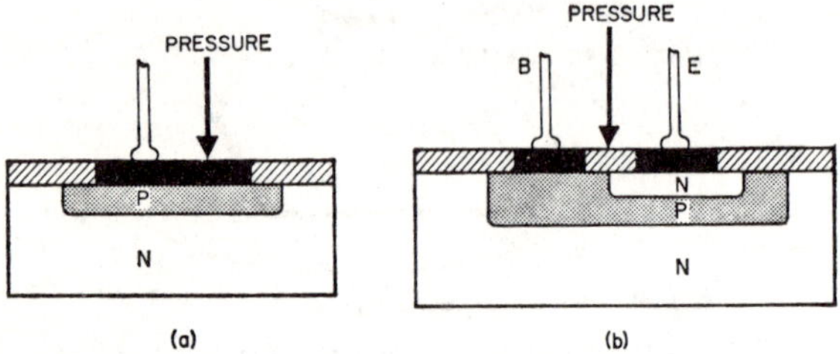

Fig. 18.6.—*Semiconductor microphone arrangements.*

a reverse voltage connected to the junction, increase of pressure will result in an increase of reverse leakage current, and so an output signal will be produced across a series load resistance.

The second method incorporates a diffused junction transistor. In this case, pressure is applied to the region of the emitter–base junction as indicated in Fig. 18.6 (b). This results in a decrease of current gain and, with a constant base input current, a signal variation corresponding to the pressure will be produced in the collector current. The transducer action will include amplification in the transistor structure, and so if a low-noise type of transistor is used the arrangement might be equivalent to a conventional microphone plus its head amplifier, thus offering a more economic system. At the present time, there are unresolved problems such as noise from the stylus pressure contact, and the method of transmitting pressure to the stylus.

APPENDIX

All transistors quoted in the practical circuits are manufactured by Texas Instruments Ltd., Bedford, England. The following tables give general characteristics. More detailed data can be obtained from the manufacturers.

Germanium transistors

Type	*Construction*	*Case*	*Limiting ratings*			*Current gain*	*Frequency cut-off* f_T (Mc/s)	*Near equivalents*
			Collector voltage (V)	*Collector current* (mA)	*Collector power* (mW)			
2G301	pnp alloy	SO2	20	200	200	30–60	7	OC41
2G309	pnp alloy	SO2	20	100	200	170 min	20	AC107
2G339	npn alloy	TO5	20	300	150	50 min	5	AC127
2G344	pnp alloy	SO2	15	100	200	30 min	12	OC44
2G345	pnp alloy	SO2	15	100	200	30 min	7	OC45
2G371	pnp alloy	SO2	20	300	200	30–135	5	OC71
2G374	pnp alloy	SO2	20	300	200	70–275	5	
2G381	pnp alloy	SO2	20	500	250	50–160	5	OC81
2G384	pnp alloy	TO5	25	1 A	250	40–150	1	ACY18
2G414	pnp alloy diffused	TO18	20	10	100	40 min	120	AF114
2N457A	pnp alloy power	TO3	60	7 A	100 Watts case at 50°C	30–90	0·4	OC29
2N514A	pnp alloy power	TO3	60	25 A	100 Watts case at 50°C	20–60	0·4	
2N1038	pnp alloy	TO5 stud	40	3 A	13 Watts case at 50°C	20–60	0·5	
2N1141	pnp diffused mesa	TO5	35	100	300	10 min	$f_\alpha = 750$	AFY19
2N1302	npn alloy	TO5	25	300	150	20–100	$f_\alpha = 5$	ASY28
2N1309	pnp alloy	TO5	30	300	150	80–160	$f_\alpha = 15$	ASY27
2N1372	pnp alloy	TO5	25	200	250	30–95	$f_\alpha = 1{\cdot}5$	ACY22
GM0290	pnp epitaxial mesa	TO18	20	50	75	20 min	700	AF139

Silicon transistors

Type	Construction	Case	Limiting ratings			Current gain	Frequency cut-off f_T (Mc/s)	Near equivalents
			Collector voltage (V)	Collector current (mA)	Collector power (mW)			
2N120	npn grown junction	Flat	45	25	150	80–300	$f_\alpha = 7$	
2N338	npn grown junction	TO5	45	20	125	45–150	$f_\alpha = 25$	
2S501	npn planar	TO18	25	30	300	40–120	30	
2N656	npn diffused mesa	TO5	60	2 A	4 Watts case at 25°C	30–90	12	
2N696	npn planar	TO5	60	500	800	20–60	40	
2N743	npn epitaxial planar	TO18	12	200	300	20–60	300	BSY27
2N929	npn planar	TO18	45	30	300	40–120	30	
2N930	npn planar	TO18	45	30	300	100–300	30	
2N1049	npn diffused mesa	Single end stud	80	500	40 Watts case at 25°C	30–90	2	
2N1131	pnp diffused mesa	TO5	50	500	800	20–45	40	
2N2411	pnp epitaxial planar	TO18	20	100	300	20–60	150	
2N2500	Diffused field effect	TO5	20	10	500	—	—	
2N2639	npn epitaxial planar	TO18	30	30	300	60 min	60	
2N997	npn epitaxial planar	TO18	40	30	300	1,000 min	—	
3N74	npn epitaxial planar	TO18	18	20	300	—	30	
TIX3016	npn epitaxial planar	Micro mesa	15	50	200	20 min	1,700	

Silicon transistors (continued)

Type	*Construction*	*Case*	*Limiting ratings* Collector voltage (V)	Collector current (mA or A)	Collector power (mW or W)	*Current gain*	*Frequency cut-off* f_T (Mc/s)	*Near equivalents*
BC182L	npn epitaxial planar	Plastic TO92	50	200 mA	300 mW	100–480	150	
BC183L	npn epitaxial planar	Plastic TO92	30	200 mA	300 mW	100–850	150	Industry
BC212L	pnp epitaxial planar	Plastic TO92	50	200 mA	300 mW	60–300	200	Standards
BC213L	pnp epitaxial planar	Plastic TO92	30	200 mA	300 mW	80–400	200	
BFR41	npn epitaxial planar	Plastic TO92	50	1 A	800 mW	100 min	100	BCX32/34
BFR81	pnp epitaxial planar	Plastic TO92	50	1 A	800 mW	100 min	100	BCX36
TIP41A	npn diffused	Plastic TO66	60	6 A	65 W	15–75	3	BD243
TIP42A	pnp diffused	Plastic TO66	60	6 A	65 W	15–75	3	BD244

Diodes

Type	***Construction***	***Case***	***Description***			***Near equivalents***
1S7047	**Alloy**	**Glass ⅛″ dia × ¼″**	**Zener diode**	**4·7 V**	**400 mW**	**Z2A47F**
1S7068	**Alloy**	**Glass ⅛″ dia × ¼″**	**Zener diode**	**6·8 V**	**400 mW**	**Z2A68F**
1S7062	**Alloy**	**Glass ⅛″ dia × ¼″**	**Zener diode**	**6·2 V**	**400 mW**	**Z2A62F**
1S5039	**Diffused**	**Top hat stud**	**Regulator diode**	**39 V**	**8 W**	—
1S101	**Diffused**	**Top hat**	**General purpose rectifier**	**200 V**	**0·75 A**	**ZR62**

INDEX

By the same author

Integrated Circuit Pocket Book

Questions and Answers on Integrated Circuits

TRANSISTOR POCKET BOOK